Worldwatch Paper 111

Empowering Development: The New Energy Equation

Nicholas Lenssen

November 1992

The Worldwatch Institute is an independent, nonprofit research organization created to analyze and to focus attention on global problems. Directed by Lester R. Brown, Worldwatch is funded by private foundations and United Nations organizations. Worldwatch papers are written for a worldwide audience of decision makers, scholars, and the general public.

Empowering Development:
The New Energy Equation

Nicholas Lenssen

Derek Denniston, Research Assistant

Ed Ayres, Editor

Worldwatch Paper 111
November 1992

Sections of this paper may be reproduced in magazines and newspapers with acknowledgment to the Worldwatch Institute. The views expressed are those of the author and do not necessarily represent those of the Worldwatch Institute and its directors, officers, or staff, or of funding organizations.

Printed on 100% recycled paper containing 15% post-consumer waste

Table of Contents

Introduction

In late 1985, the Brazilian government added a small new office to its mammoth national electric utility. The new National Electricity Conservation Program (*Programa Nacional de Consevação de Energia Elétrica*, or *PROCEL*) was given a budget as tiny as its name was long.

Over the ensuing four years, *PROCEL* invested a modest $20 million in energy-efficient refrigerators, electric motors, and other power-saving technologies, and in helping private businesses bring them to market. It replaced some 280,000 inefficient street lights with new mercury vapor and high-pressure sodium lights that reduced electricity use by up to 70 percent. Its investments were matched by a similar outlay from local governments and industry. By 1990, the conservation program had saved enough electricity to reduce the nation's need for new power plants and transmission lines by roughly $1 billion. For each $1 invested, $25 was freed up for other, urgently needed uses.[1]

PROCEL's approach, which emphasizes maximizing energy *services*, represents a major break from the traditional emphasis on simply increasing gross *supplies* of oil, coal, or electricity. Instead of huge investments in new power plants, it utilizes smaller investments in more efficient use of already-existing power supplies—producing more service at much lower cost. The new energy equation has not only impelled Brazilian authorities to reassess some of their plans for expensive new energy projects; it is provoking similar reassessments in China, Thailand, and other developing countries. It has aroused interest not just because it is a way of reducing the burdens of financially strapped utilities, but because it offers a means of revitalizing stalled development efforts throughout much of the world.

I am grateful to Ashok Gadgil, Stephen Karekezi, and Marnie Stetson, as well as to my colleagues at Worldwatch Institute, for reviewing drafts of this paper.

6 Energy has long been recognized as a key to development for the 4 billion people who live in Africa, Asia, and Latin America. Among planning and funding institutions, a virtually unquestioned assumption has been that an expanding energy supply is the necessary foundation for expanding industries, providing jobs, and raising standards of living in the developing world (or "South"). That assumption has seemed logical enough, since high energy use is a primary characteristic of the most highly developed nations.

On the basis of that assumption, developing countries have more than quadrupled their energy use since 1960. Per-capita use has more than doubled. Yet the strategies that have been so successful in achieving this growth have left these nations staggering from oil price shocks, struggling with foreign debt, and suffering from serious environmental and health problems—while still facing severe energy shortages.[2]

Despite the rapid rise in Third World energy use, the income gap between North and South has been widening. In 1960, the people in the richest fifth of the world's countries received 30 times more income per person than those in the poorest fifth, according to the United Nations Develment Programme; yet by 1989, despite more rapid energy development in the South than in the North, the disparity had widened to 60-to-1. Over the past decade, per capita incomes declined in some 50 countries; and in Latin America, where some of the world's largest energy projects have been built, three-quarters of the population saw its incomes fall. The poor have benefited little from rising energy use.[3]

Incomes declined in the eighties partly because developing economies were saddled with back-breaking debts to foreign banks and governments—debts totalling $1.35 trillion in early 1992. And as debt deepened poverty, rising energy use deepened the debt. In Brazil and Costa Rica, for example, one of every four borrowed dollars went to pay for giant electric power projects.[4]

Living standards were then further battered by the strategies used to pay off the debt. In many countries, much of the income required to pay off foreign debts came from the exports of energy-intensive products, such as aluminum. Not only was a large portion of national income thus diverted from badly needed social services such as education and health, but much of the energy itself was being diverted to subsidize those export products.

Further undermining of development occurred as the big energy projects caused devastating environmental damage, from erosion and acid-rain destruction of cropland to urban pollution—all of which directly diminished living standards. Perhaps most alarmingly, conventional energy development has increased the threat of global warming. Although industrial countries have emitted 79 percent of the fossil fuel-derived carbon dioxide released since 1950, and still accounted for 69 percent of the total in 1990, future growth is expected to come more from the South than from the North. And while some industrial-nation politicians accuse developing countries of threatening the global environment, the South tends to reject this as a reason to alter its energy strategy; its argument is that since carbon dioxide remains in the atmosphere for well over 100 years, it is largely the North that caused the problem.[5]

The salient question, though, is not whether developing countries have the right to follow the same energy path as the North, but whether it's in their own interest to do so. The promise of development based on heavy industry, often fueled by coal, is a mirage that has been pursued with a notable lack of success by the Soviet bloc countries in the past, and by China and India more recently. Industrial facilities in the developing countries, like those in the former Communist world, are less productive than factories in the North, but use more energy and emit more pollution.[6]

This well-worn development path has proven an economic and environmental disaster. If developing countries are to achieve the hoped-for gains in living standards, they need to meet their energy needs in a way that allows them to close the gap between North and South instead of falling further behind. Yet it is now clear that they cannot hope to do this simply by expanding energy supplies as they have in the past.

The way out of this dilemma is to focus on providing the energy services, such as cooking, lighting, and increased agricultural productivity, that are the real keys to economic and social development. As Brazil has already demonstrated on a small scale, investing in conventional energy supplies is not always the best way to meet energy needs, though power plant builders and development officials continue to act as if they believe otherwise.

Developing countries can accomplish this change by following a two-part strategy: first emphasizing the use of more energy-efficient technologies in everything from industrial processes to consumer products; and second, meeting the remaining need for new energy supplies by developing far less costly and ecologically destructive resources than those they have pursued to date. With this strategy, developing countries can "leapfrog" to the advanced technologies being commercialized in industrial countries today, avoiding billions of dollars of misdirected investments in infrastructure that is economically and environmentally obsolete.[7]

Over the next 35 years, $350 billion invested in efficiency improvements could eliminate the need for $1.75 *trillion* worth of power plants, oil refineries, and other energy infrastructure—opening the way for vastly larger investments in food production, health, education, and other neglected needs. A concerted move toward efficiency would also lead to greater employment—a major benefit in Third World economies that are labor-rich and capital-poor. Finally, and of critical importance to the health of the global environment, an efficient energy system would delay global warming by slowing down the increase in carbon emissions.[8]

Such a change in investment priorities will require major reforms—both in developing-country institutions and in the industrial countries that control international financial organizations such as the World Bank. Yet the alternative—continuing capital shortfalls, repeated oil shocks, and ever-growing emissions of global warming gases and air pollutants—would only lead to stagnant development prospects for most of the people on the planet.

The Energy Dilemma

Rapid industrialization and urbanization have led to large increases in developing countries' use of oil, coal, and other sources of energy. Since 1970 alone, energy consumption in these nations has nearly tripled—a rate of increase fifteen times that of the industrial countries. It has grown far faster than population, faster even than their economies. (See Figure 1.) By 1991, developing countries (home to 77 percent of the world's people) used 24 percent of the world's oil, coal, natural gas, and electricity. Adding wood and agricultural wastes, which provide more than a third of the energy in these nations, brings the total to some 30 percent of the world's energy use.[9]

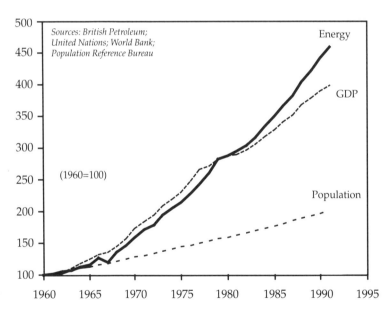

Figure 1: Growth in Developing Countries:
Energy, GDP, and Population, 1960–1991

Despite this increase, people in developing countries still use just one-ninth as much commercial energy on average as those in industrial countries do. (See Table 1.) And wide disparities exist: Bangladeshis use one-tenth as much as Thais, while Thais use one-third as much as South

Table 1: World Commercial Energy Consumption, 1970 and 1990

Region	1970 Energy Consumption (exajoules)	1970 Per Capita (gigajoules per person)	1990 Energy Consumption (exajoules)	1990 Per Capita (gigajoules per person)
Developing Countries	30	12	84	21
Latin America	8	26	16	37
Asia	19	10	59	20
Africa	4	10	9	14
Centrally Planned Economies	44	120	71	167
Industrial Countries	129	180	154	185
WORLD	203	55	310	59

Sources: British Petroleum, *BP Statistical Review of World Energy* (London: 1992); United Nations, *World Population Prospects 1990* (New York: 1991).

Koreans. The differences within countries are similar to those between them, with elite minorities devouring as much energy as Americans or Europeans, while most rural dwellers use little at all.[10]

Per-capita consumption, though, can be a misleading indicator of economic or social well-being. The *service* that energy provides is important; the actual *amount* used is not. While the typical Indian uses slightly more energy than a neighboring Pakistani, for example, the Indian's income gives him only half as much purchasing power as the Pakistani's. Similarly, in 1987, the average East German consumed 41 percent more energy than a West German, though the West German was nearly twice as wealthy. Indians and East Germans were thus not getting as much service for each barrel of oil or kilowatt-hour of electricity as their neighbors were.[11]

Energy priorities in developing countries have been dictated largely by the appetites of industrial countries. They are particularly affected by the demand for oil, which has led multinational companies to search the globe for new supplies, and to encourage developing countries to export it for foreign exchange. Due to the ease with which oil can be moved and stored, most developing countries have also followed the practice of Northern countries in using oil to fuel industry and transportation, as well as for cooking, lighting, and producing electricity. But while a few nations have improved their balance sheets through oil exports, most face a continual drain on their economies as a result of this dependency. Scarce foreign exchange earned through exports of agricultural products or minerals is spent to import oil for domestic consumption, draining resources away from development. Three-fourths of developing countries are oil importers. And of the 38 poorest countries, 29 import more than 70 percent of their commercial energy—nearly all of it in the form of oil.[12]

Oil importers are highly vulnerable to the rapid price increases and energy shortages that have rocked the global economy three times in the past 20 years. These shocks left economies in both industrial and developing countries in economic recessions, but developing countries suffered disproportionately more. One reason was that recession in industrial countries decreased demand for Southern export products. As a result, at the same time that developing countries were being hit with higher bills for imported oil, they were getting reduced earnings for their exports. Perhaps hardest hit were the countries of sub-Saharan Africa. Following the 1973 Arab oil embargo, oil import bills there went from absorbing 10 percent of export earnings to absorbing more than 20 percent. After the Iranian revolution of 1979, oil imports gobbled more than half the export earnings of many poorer countries. Even at today's lower oil prices, nearly a third of sub-Saharan Africa's hard-currency earnings are spent on petroleum imports, effectively gutting investment in other areas.[13]

The use of valuable foreign exchange to import energy means that foreign debt loads, too, get bogged down by oil. During the past decade, India's oil import bill soaked up nearly a third of export receipts and totalled $36.8 billion—equivalent to almost 87 percent of its new debt, according to the International Energy Initiative in Bangalore. With foreign purchases of electric generating equipment included, foreign energy expenses accounted for more than four-fifths of India's export earnings between 1980 and 1986. When oil prices skyrocketed in late 1990 following the Iraqi invasion of Kuwait, India was suddenly caught with oil bills larger than the export income it depended on to pay them. The country was forced to ration gasoline and sell overseas securities to avoid defaulting on its debt. Consequently, in 1991, India had its lowest economic growth rate in over a decade.[14]

In the developing world, even some oil exporters have seen their ability to maintain foreign sales eroding as domestic consumption rises. Indonesia, for example, now consumes 44 percent of the oil it pumps, up from 26 percent in 1981—and could become a net oil importer as early as 1997. Likewise, China's domestic consumption has risen faster than its output, despite efforts by the government to discourage oil use in order to maintain exports. The result has been a 70 percent cut in proceeds from international oil sales since 1984. Exports could completely cease by 1995.[15]

Energy troubles in developing countries, though, are not limited to volatile oil prices and growing supply shortages. The foreign debt loads that burden these countries originate partly in massive borrowing to pay for energy infrastructure. As a result, in much of the Third World, government-owned power companies are deeply in debt from electric power construction programs. In the eighties, a quarter of the money developing country governments paid to Northern creditors went to pay off past energy projects. (With the onset of the Third World debt crisis, the percentage of new borrowing used for expanding electricity supplies declined by half, but the millstone effect of powerplant investment had been amply demonstrated.) Much of the money—in Latin America, 70 percent—was borrowed from commercial banks, at higher interest rates than those charged by the World Bank and similar institutions. At the same time, in efforts to boost economic growth, stem inflation, or simply win popular support, governments slashed electricity prices, with tariffs dropping from an average of 5.5 cents a kilowatt-hour in 1983 to 3.8 cents by 1988. As a result, consumers in developing countries now pay just 60 percent of the cost of producing electricity. In short, many utilities are not earning enough money even to cover their monthly bills, much less pay back foreign banks.[16]

Yet many developing countries, despite their heavy investments, still face shortages of electricity. India's shortfall averages 9 percent, rises to 22 percent during peak periods, and is worsening. China's shortfall results in regular shutdowns of industry; it idled one-fourth of the country's industrial capacity in 1987. And in Latin America, electricity shortages are costing industry as much as $15 billion per year in lost output.[17]

13

Such shortfalls have social repercussions, as well. In Calcutta, citizens attacked utility workers following a power outage in 1991. Bombay has been on the verge of riots over breakdowns twice in the past three years. India's then secretary to the department of power, S. Rajgopal, declared that "power is no longer just an economic issue. It's a question of law and order."[18]

To close the generating gap, electric utilities throughout the developing world are building power plants as fast as they can. Present plans call for spending some $100 billion—including $40 billion in foreign exchange—on new power plants and transmission lines each year through the nineties, according to the World Bank. For many utilities, however, implementing these plans will be impossible. The inability of utilities to repay their existing debt has reduced the willingness of private banks to lend them more money. Third World utilities may have difficulty borrowing even half the $40 billion a year the World Bank says is needed. And internal capital markets are unlikely to make up the rest, given current risk levels. Senior World Bank economist John Besant-Jones foresees massive economic disruption in developing countries during the nineties if power sector crises go unresolved.[19]

Cash-strapped utilities that cannot meet present obligations have little prospect of extending power supplies to areas still without electricity. Some 2.1 billion people in the developing world still live in such areas. For most, there is little prospect of that changing anytime soon. The traditional means of providing power to rural communities—extending power lines throughout a country—is too expensive for many governments to keep subsidizing. Building lines costs up to $15,000 per kilometer. In most countries, utilities manage to hook up only 2 to 3 percent of unserved rural families to the electric grid each year—a rate that in many areas is not even fast enough to keep pace with population growth, according to London-based energy consultant Gerald Foley.[20]

14 Rural energy users, while deprived of wired electricity from distant sources, must also cope with a scarcity in nearby biomass resources such as wood. At least 2 billion people depend almost exclusively on biomass for their energy supplies, though their total energy consumption is relatively small in global terms (if they could afford to burn petroleum or bottled gas to meet their current energy use, they would produce an 8 percent increase in world oil consumption). The U.N. Food and Agriculture Organization, meanwhile, estimates that the number of people suffering from wood fuel shortages will grow from 100 million in 1980 to more than 350 million by the end of the decade.[21]

While development experts no longer consider fuelwood collection to be the driving force behind most deforestation (land clearance for agriculture ranks as the primary cause), scouring the countryside for fuelwood places severe pressures on the land—and on people. Shortfalls in fuelwood contribute to increased fuel costs for urban biomass users, longer collection times for women and children in the countryside, and a reduction of crop residues and dung returned to agricultural land when these natural fertilizers are used as a replacement fuel. The inevitable result of higher fuel prices, or shortfalls in traditional supplies, is a worsening not only of environmental degradation but of poverty.[22]

Over the long run, countries need healthy people, forests, cropland, and waterways to maintain a productive economy; yet environmental and health costs associated with energy use and production are taking a growing toll in developing countries. These costs pose a worldwide threat to developmental goals. Burning wood in traditional stoves, for instance, pollutes the air in Third World kitchens. Some 400 to 700 million people, primarily women and children, suffer from the carbon monoxide, particulates, and cancer-causing chemicals such as benzopyrene emitted by these stoves. Acute respiratory infections are the leading killer of children under five, accounting for more than 3.5 million deaths annually, according to the World Health Organization. Smoke from cooking fires contributes to widespread respiratory problems among women as well.[23]

Urban areas are beset with energy-related air pollution problems, primarily from motor vehicles. Mexico City, with its 20 million residents, has become the *ne plus ultra* of uncontrolled air pollution; ozone levels there violated international standards 303 days in 1990, up from 130 days in 1986. When record air pollution levels were reached in early 1992, Mexican officials ordered more than 400,000 cars off the road, cut industrial production by 75 percent, and told schoolchildren to remain at home. Mexico City is joined by Bangkok, Nairobi, Santiago, and São Paulo in the growing list of cities whose people suffer from lung-damaging air pollution, *despite* per capita levels of energy consumption far below those of Northern cities. China, the world's largest consumer of coal, has seen sulfur dioxide concentrations regularly surpass international guidelines in major cities such as Beijing. The level of suspended particles in Chinese cities is 14 times that in the United States. The result is a growing public susceptibility to eye irritations, skin reactions, lung disorders, and heart ailments.[24]

Oil and coal burning can result in wide-ranging emissions of sulfur and nitrogen oxides, much of which ends up in precipitation. In China, acid rain falls on at least 14 percent of the country, damaging forests, crops, and water ecosystems. In the Philippines, pollution from a coal plant in Batangas province is reported to have reduced output of rice and sugar, while increasing respiratory problems among the residents. As in industrial countries, public protest over environmental effects of energy use has hindered plans to expand supplies. The reported damage in Batangas, for instance, has stalled construction of other Philippine power plants.[25]

Industrial countries have recently shown greater interest in Third World energy use, partly because of growing recognition that greenhouse gases—from whatever sources—pose a threat to the world as a whole. In its 1992 report to the U.N. Conference on Environment and Development (UNCED), the Intergovernmental Panel on Climate Change reiterated its earlier finding that doubling atmospheric carbon dioxide—generated mainly by the burning of fossil fuels—will warm the planet by 1.5 to 4.5 degrees Celsius by the end of the next century. If recent trends continue, developing-world emissions will increase from 1.8 billion tons of carbon in 1990 to 5.5 billion tons in 2025. China alone is forecast to emit more carbon dioxide by 2025 than the current combined total of the United States, Japan, and Canada. Such increases would boost global emissions by half at a time when they should be *reduced* at least 60 percent if the atmospheric concentration is to be stabilized and global warming minimized. A reduction of that magnitude will require major changes in all countries—industrial and developing.[26]

At present, most developing-country governments accept increasing emissions of carbon dioxide as a necessary evil if they are to raise living standards. Yet they are also aware of the risks they run if predictions of climate change come true. Sea level rise, for example, could inundate 33 island nations, as well as low-lying coastal regions of populous countries such as Bangladesh and Egypt—leading to massive dislocations. Research in Indonesia, Kenya, and Malaysia confirms that agricultural output would decline as a result of global warming. And China's meteorological bureau is concerned that global warming already may be exacerbating a severe drought in the northern part of the country.[27]

Mitigating global warming and other environmental impacts will be difficult at best, given that future population growth alone will spur a 70-percent jump in global energy use in 30 years even if per capita consumption remains at current levels. With high rates of economic growth, these nations could triple their energy use by 2020, according to conventional projections. To achieve this by following the established path would require a tripling of developing world oil, coal, and natural gas use—and the building of thousands of new power plants. Many expert observers no longer find such plans realistic, given the economic and environmental constraints.[28]

Yet the threat of climate change might be turned to the advantage of developing countries if their industrial neighbors, seeking to slow global warming, help the Third World build efficient systems that limit greenhouse gas emissions while meeting energy service needs. Indeed, it now seems likely that the planet's atmosphere can only be conserved by ensuring that every country charts a sustainable energy path.

Efficiency: The Coming Revolution

Since the 1973 oil embargo, industrial countries have made large gains in using energy more economically. For the 24 member nations of the Organisation for Economic Co-operation and Development (OECD), energy use rose only one-fifth as much as economic growth between 1973 and 1989. However, these gains have largely bypassed developing countries, where energy use expanded 20 percent faster than economic growth during the same period.[29]

Developing-country economies now require 40 percent more energy than industrial ones to produce the same value of goods and services. Part of the difference is due to the fact that these countries are still building energy-intensive infrastructure and related industries—but often using outdated technologies that squander energy. The gross inefficiencies of these technologies—whether in wood stoves, cement plants, light bulbs, or trucks—offer innumerable opportunities to limit energy consumption and expenditures while expanding the services they provide. For example, the U.S. Office of Technology Assessment estimates that nearly half of overall electricity use in the South can be cut cost-effectively.[30]

To compete in increasingly integrated world markets, while still meeting their own domestic needs, developing countries will need to reap the economic savings that improved energy efficiency offers. But it will take a concerted effort by consumers, businesses, and governments to capture the full potential. Individual consumers, particularly poorer ones, often cannot afford to buy more efficient appliances, and have no incentive to do so under government policies which subsidize energy but not efficiency. Manufacturers and importers lack incentives to reduce products' energy use even when no additional cost is involved. Meanwhile, governments and international lending agencies usually direct their money and efforts toward simply expanding supplies, while paying little attention to how much heat or power—and how much pollution—is produced per investment dollar.

Half of Third World commercial energy consumption goes to industry (See Table 2), yet for each ton of steel or cement produced, the typical factory in the global South uses more energy than its Northern counterpart. Steel plants in developing countries, for example, consume roughly one-quarter more energy than the average plant in the United States, and about three-quarters more than the most efficient plant. Fertilizer plants in India use about twice as much oil to produce a ton of ammonia as a typical British plant does. Pulp and paper facilities consume as much as

Table 2: Energy Use in Developing Countries, by Sector, 1985

Sector	Commercial[1]	Biomass	Total
	(percent)		
Residential and Commercial	28	90	44
Industrial[2]	52	10	41
Transportation	20	—	15

[1]Commercial energy refers to coal, oil, natural gas, electricity and other fuels that are widely traded in organized markets.
[2]Includes agriculture.

Source: U.S. Congress, Office of Technology Assessment, *Fueling Development* (Washington, D.C.: U.S. Government Printing Office, 1992).

three times more energy for the same amount of output. Such records are often the result of poor maintenance and operating procedures, and can be readily improved—given sufficient information and incentive to do so. Indonesian industries, for example, could cut energy use 11 percent without *any* capital investment, simply by changing operating procedures. Similarly, a Ghanaian survey found potential savings of at least 30 percent in medium- to large-scale industries.[31]

Efficiency is also impaired by reliance on old or obsolete industrial processes—often purchased at bargain prices from Northern countries. Cement plants operate in 84 of 110 developing countries, often ranking as the most energy-consuming industry; these plants typically use 50 to100 percent more energy than the best ones in industrial countries, partially due to reliance on an antiquated wet process. In Kenya, more than two-thirds of the country's industrial energy goes to making cement. In Tunisia, a government program prompted the modernizing of cement-making in the early eighties, and the energy efficiency of the industry was improved by 13 percent over eight years.[32]

Some of the largest opportunities to save energy and money are in the electric power industry. Third World power plants typically burn one-fifth to two-fifths more fuel for each kilowatt-hour generated than those in the North, and they experience far more unplanned shutdowns for repairs, as they are often poorly maintained and operated by inadequately trained staff. Once electricity is generated, 15 to 20 percent of it disappears in line losses and theft, as industries and individuals hook into power lines without paying for the service. In some countries, the rates are even higher: Bangladesh reportedly loses over 40 percent of its generated power this way.[33]

19

Because the developing world is still in the early stages of building its industrial infrastructure, it has opportunities to base future development not just on more efficient processes, but also on more efficient products. Building a $7.5 million compact fluorescent light bulb factory, for example, would eliminate the need to build $5.6 *billion* worth of coal-fired power plants, if the bulbs (which need 75 percent less power than incandescent ones) were used domestically. Each dollar invested in efficiency would save $740 in capital expenditure—before the savings on energy use even begin.[34]

In Pakistan, a switch from incandescent to compact fluorescent bulbs helped Karachi's Aga Khan Hospital cut its total energy consumption 20 percent. A $10 million factory making advanced windows for commercial buildings would offset $4.6 billion worth of power plants. Such a strategy would also generate products for which there is high international demand. With compact fluorescent bulbs, which are already manufactured in Brazil, Mexico, Taiwan, China, and Sri Lanka, global demand grew by 36 percent per year over the past two years. In North America, Europe and Asia, demand now exceeds supply.[35]

The emphasis on energy-intensive industrialization typically found in developing countries can have unforeseen impacts on economies already burdened by underemployment and debt. In Brazil, for instance, the government subsidizes electricity for energy-intensive industries such as aluminum smelting (which pays just one-third the actual cost of producing power) to boost exports that can service its foreign debt. As a result, the government has been stuck with the enormous costs of building hydroelectric dams, which supply power to the smelters and other plants. A move toward lighter industries, such as computers, could create 120 times as many jobs and generate 20 times as much tax revenue as a similar investment in energy-intensive, export-oriented industries such as aluminum. A similar situation exists in Karnataka, India, where metal-producing plants (primarily steel and aluminum) use 69 percent of the industrial power consumption, yet provide only 9 percent of industrial employment. Indeed, some economists question the whole notion of economic development based on resource- and energy-intensive industrialization.[36]

In agriculture, too, there is an urgent need for movement toward greater energy efficiency, particularly in view of the fact that agricultural inputs in the Third World—including fertilizers, pesticides, irrigation, tractors, and processing—are expected to become more energy-intensive as population growth drives up the demand for food. In China, chemical fertilizer use has gone up 140 percent since the late seventies, while grain production has increased only 26 percent. Still further increases in fertilizer use are expected.[37]

By carefully selecting the technology they use, developing countries can increase agricultural output while keeping expensive, energy-intensive inputs to a minimum. In China and India, for example, the ratio of nitrogen fertilizer to potassium and phosphorus fertilizers far exceeds that found in other countries, suggesting that much of the nitrogen fertilizer is not being optimally incorporated by plants. In India, researchers have developed a form of nitrogen fertilizer that provides better delivery to the crop roots, allowing a 35 percent reduction in fertilizer applied. And in the United States, researchers have shown that mechanical tillage in corn and soybeans is less expensive and energy-intensive (while more labor-intensive) than chemical weed control.[38]

Agriculture is energy-intensive not only because of its demand for chemicals, but because it is—in some developing countries—a large consumer of electricity. India's 8 million irrigation pumps, which use nearly one quarter of the country's electricity, employ inefficient motors and poorly designed belts, and are plagued by leaky foot valves and high friction losses. Using more efficient pumps could cut electricity consumption by roughly half, at a cost of only 1 cent per kilowatt-hour saved. Unfortunately, tariffs for electricity are generally so low that

> **"A typical Chinese refrigerator uses 365 kilowatt-hours of electricity per year, whereas a similarly sized Danish one needs less than 100 kilowatt-hours."**

farmers have no incentive to conserve energy. Even so, one retrofit program conducted by the Indian Rural Electrification Corporation in the mid-eighties reduced electricity consumption in 23,000 pumps by a quarter, and the improvements paid for themselves in less than six months. Similar results have been achieved in Pakistan, which also relies heavily on electric pumps for irrigation.[39]

Although industry and agriculture still consume most of the commercial energy in developing countries, the urban residential and commercial sectors are growing much faster. In Thailand, some 40 percent of the projected increase in electricity growth is for commercial buildings. Heating water and cooking are still the primary energy uses in urban and rural households, but most of the growth is in electricity-consuming products such as lights, televisions, refrigerators, and even air conditioners. In China, only 3 percent of Beijing's households had refrigerators in 1982; six years later, 81 percent did. Lighting alone accounts for one-third of India's peak electricity demand, and one-sixth of its total electricity consumption.[40]

For consumer products, as for industrial processes, developing countries often rely on outdated technologies that consume more energy than needed. A typical Chinese refrigerator, for example, uses 365 kilowatt-hours of electricity per year, whereas a similarly sized South Korean model uses 240 kilowatt-hours, and a Danish one needs less than 100 kilowatt-hours. Yet industrial planners and manufacturers in developing countries are rarely concerned with the energy efficiency of their products—only with producing and selling more of them by keeping the initial cost as low as possible.[41]

The same can be said of architects and civil engineers. Much of the developing world relies on air conditioning in commercial buildings. Improved building designs—including insulation, better windows, and natural ventilation—could cut cooling needs and costs, but such designs are not widely used. In Bangkok, for example, large offices typically use windows made of a single sheet of glass. By substituting advanced double-paned windows with a special low-emissivity coating (which filters out infrared rays but allows in visible light), builders would reduce not only the subsequent electricity costs but the initial costs of construction, since they would then be able to install smaller, less expensive air conditioning units. And while it is preferable to make such substitutions during construction—when developing countries are building their infrastructure—it is also possible to upgrade existing buildings.[42]

Chinese buildings use three times as much energy for heating as comparable U.S. buildings, even though inside temperatures remain colder. By making boiler improvements and using insulation and double-glazed windows, the Chinese could raise average building temperatures from 11 degrees Celsius to 18 degrees—while consuming 40 percent less coal. One study found that such improvements can pay for themselves in 6.5 years in the northern city of Harbin even with subsidized coal; with unsubsidized coal, the payback would be around four years.[43]

Efficiency improvements can even be made in the use of biomass—wood, charcoal, or agricultural residues—for cooking, allowing women to spend less time or money acquiring fuel. Traditional Third World cooking stoves operate at an efficiency level of around 10 percent, but improved stoves can convert 20 to 30 percent of the fuel to useful cooking energy. Similar gains in efficiency can be achieved in converting wood to charcoal for urban markets or rural industries.[44]

In the seventies, development organizations put a high priority on improving cooking stoves, but because of problems with reliability and affordability the new stoves did not "catch on" at first. Now that situation is changing. In Kenya, an improved charcoal stove—the ceramic jiko—has become a major success. More than half a million have been sold, with at least 130,000 being added each year. Kenya's success has helped to inspire similar programs in 15 other countries in Africa, with some 150,000 improved stoves sold in Niger and 200,000 in Burkina Faso. In India, a government program had distributed some six million advanced cookstoves by early 1989.[45]

A third venue of rapidly growing energy consumption in developing countries is transportation. In China, transportation doubled its percentage of national oil consumption between 1980 and 1988. In most developing countries, this sector accounts for over one-half of total oil consumption, and one-third of commercial energy use. Much of the increase has been caused by the rapid growth in urbanization and ownership of cars and two- or three-wheeled motor vehicles. During the past decade, car registrations shot up in Asian developing countries by over 10 percent annually. (See Table 3.) Meanwhile, two- and three-wheeled motor vehicles outnumber automobiles in many Asian cities.[46]

Table 3:   Registered Automobiles, Selected Countries and Regions, 1970, 1980 and 1990

Country/ Region	1990 Registered Automobiles	1970	1980	1990
		People per Automobile		
	(thousands)			
Asia	23,195	517	290	121
Africa	8,821	104	72	75
Latin America	30,006	37	19	15
Western Europe	146,582	5	3	3
United States	143,550	2	2	2
WORLD	444,900	18	14	12

Source: Motor Vehicle Manufacturers Association, *World Motor Vehicle Data* (Detroit: various years).

Motor vehicles made in developing countries often fall below the efficiency levels found in the North. New cars manufactured in Brazil, for instance, are 20 to 30 percent less efficient than comparable models made in Europe and Japan. Meanwhile, prototype vehicles built in the industrial world can quadruple the current level of efficiency. For buses and trucks, which in developing countries consume more petroleum overall than private automobiles, gains are also feasible, though less dramatic. By switching to improved engines that are domestically produced, new Indian buses can be made up to 20 percent more efficient. The new engine pays for itself in saved fuel in three years. Further efficiency gains can be achieved through more extensive training of drivers and vehicle maintenance workers.[47]

Unfortunately, the growing automotive congestion in many Third World cities can negate even these vast gains. Congestion in Bangkok has dropped the average vehicle speed from 12 kilometers per hour in 1980 to an estimated 5 kilometers per hour today. More serious measures are needed to improve transportation of people and goods, including improved traffic management, mass transportation, and better planning for future land use activities. In particular, promoting the continued large-scale use of bicycles and other nonmotorized vehicles would reduce future motorized vehicle energy use, air and noise pollution, and traffic congestion, while boosting employment and mobility of the poor.[48]

Fully occupied buses, for example, use as little as one-sixth as much energy per passenger as automobiles. Curitiba, Brazil relies on a network of feeder and express buses that utilize dedicated traffic lanes, providing riders with fast and convenient trips, with minimal petroleum use. While Curitiba has the highest car ownership of eight Brazilian cities compared in a recent study, its bus system gives it the lowest per capita fuel consumption. Singapore, similarly, has made transportation efficiency a priority in its urban planning, taxing both ownership of cars and use of roads to increase reliance on mass transit.[49]

Outside of urban areas, encouraging the use of railways and barges in place of trucks can save both money and oil. In India, rail transport is six times more efficient than shipping by truck, yet rail's share of freight has fallen from 88 percent in 1950 to 46 percent in 1989. A reversal of this trend would save the country foreign exchange by reducing oil imports.[50]

In combination, the efficiency potentials now within reach for industry, agriculture, buildings, and transportation could provide an enormous boost to the economies of developing countries. By investing $10 billion a year, these countries could cut future growth of their energy demand by half, lighten the burden of pollution on their environments and health, and staunch the flow of export earnings into fuel purchases. Gross savings would average $53 billion a year for 35 years, according to a study prepared by scientists at the U.S. government's Lawrence Berkeley Laboratory for the U.S. Working Group on Global Energy Efficiency.[51]

"By investing $10 billion a year, developing countries could cut future growth of their energy demand by $53 billion a year for 35 years."

Although such wide-scale savings remain paper prophecies, some countries have had notable successes. In 1980, China launched an ambitious efficiency program to improve energy use in major industries. By directing roughly 10 percent of its energy investment to efficiency over five years, the nation cut its annual growth in overall energy use from 7 percent to 4 percent, without slowing growth in industrial production. Efficiency improvements accounted for more than 90 percent of the energy savings, with shifts toward less energy-intensive industries yielding the remainder. And efficiency gains were found to be one-third less expensive than comparable investments in coal supplies. One result was that China's energy consumption expanded at less than half the rate of economic growth from 1980 through 1988.[52]

China achieved this result in the midst of building its infrastructure, a phase of industrialization that in most countries has involved soaring energy consumption. Had the nation failed to make such progress, energy consumption in 1990 would have been 50 percent higher than it actually was. (See Figure 2.) Unfortunately, China has poured money into expanding its energy supply since the mid-eighties, while spending on efficiency has declined to just 6 percent of total investment in the energy sector.[53]

Brazil's National Electricity Conservation Program (PROCEL), as noted earlier, has catalyzed impressive savings of energy and money. The $20 million it spent over four years was spread over more than 150 efficiency projects and programs, with local governments and private industry providing matching funds. Most of the money went to information, education, and promotion programs to increase awareness of the savings efficiency could generate. It also encouraged the National Development Bank to offer low-interest loans to businesses willing to invest in efficiency. These efforts yielded electricity savings worth between $600 million and $1.3 billion in reduced need for power plants and transmission lines. By making broader investments, Brazil could cost-effectively eliminate 42 percent of its projected growth in electricity consumption by 2010, estimates Howard Geller, executive director of the American Council for an Energy-Efficient Economy, who has intensively studied the Brazilian energy sector. This reduction would come from improvements in just 20 areas, including refrigeration, electric motors, and air conditioning.[54]

Exajoules, commercial energy

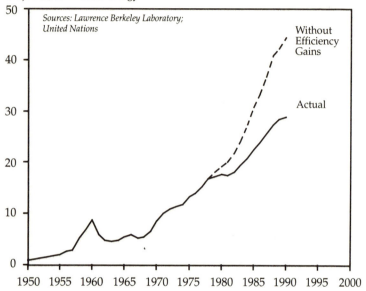

Figure 2: Energy Consumption in China, Actual
and without Efficiency Gains, 1950–1990

Brazil and China need not be anomalies, as similar potential exists throughout the developing world. Halving the rate at which the developing world's energy demand grows over the next 30 years would hold the overall increase to a doubling rather than tripling of consumption. That difference could have incalculable consequences for environmental and human health worldwide—and for the ability of the developing world to meet the basic needs of its growing population.

New Supply Choices

If developing countries proceed to squeeze all the waste they can out of the way they use energy, they will greatly reduce—but not eliminate— the need for increased supplies. In the long run, it will still be necessary to develop new energy sources. Unfortunately, government planners and international institutions still assume that developing countries have to follow the energy path the North blazed a century ago—a strategy that relies primarily on expanding supplies of coal and oil. These two fossil fuels already provide 51 percent of all energy used in developing countries (see Table 4), and more than 75 percent of commercial energy.[55]

Over the long haul, the Third World will need to develop its own alternatives to costly oil and polluting coal. Many developing countries have extensive, unexploited reserves of natural gas, which could supplant oil and coal use in buildings, transport, industry, and power generation. And all have enormous potential to rely on solar, wind, biomass, or geothermal energy resources. Driven by technological advances and cost reductions, western countries are increasingly pursuing these options. In developing countries the opportunities are even greater. Decisions they make today will determine how readily they tap these resources in the future.

Oil still provides most commercial energy for developing countries, and its use continues to grow. During the past decade, a period when oil con-

Table 4: Energy Supplies in Developing Countries, 1991[1]

Source	Share
	(percent)
Biomass	35
Oil	26
Coal	25
Natural Gas	8
Renewables	6
Nuclear	<1

[1]Primary energy.

Source: Worldwatch Institute based on British Petroleum, *BP Statistical Review of World Energy* (London: 1992), United Nations, *1990 Energy Statistics Yearbook* (New York: 1992), and J.M.O. Scurlock & D.O. Hall, "The Contribution of Biomass to Global Energy Use," *Biomass*, No. 21, 1990.

sumption declined in North America, Europe, and the former Soviet Union, it increased by a third in the developing world. As world oil reserves and production become more concentrated in the volatile Persian Gulf region (now the site of nearly two-thirds of the world's proven oil reserves and more than a quarter of its production), the likelihood of disruptive oil cutoffs or rapid price rises will grow.[56]

Other energy resources, notably coal and hydroelectric power, face growing problems of their own, especially since most of the less expensive resources have already been exploited. Coal resources are concentrated in a few countries, including China, India, South Africa, Indonesia, Colombia, Botswana, and Zimbabwe, so most nations would remain energy importers even if they chose to use coal. Even the big coal users, China and India, face obstacles. In China, shipping coal from mines in the north to the eastern economic heartland has led to transport gridlock and supply shortfalls; coal shipments already account for 40 percent of railway tonnage. In India, where most coal is of poor quality, public opposition to strip mines and polluting power plants has escalated in recent years. Both countries will need to move toward more efficient, less polluting methods to convert coal to electricity and heat.[57]

Energy planners hold high expectations for hydroelectric power, which currently provides a third of developing-country electricity. Less than 10 percent of its technical potential has been tapped. Yet exploiting the rest has run into roadblocks and declining orders in recent years as the real costs of building large dams have become more apparent. High capital costs put electric utilities deep in debt. (Brazil's and Paraguay's Itaipu Dam, for example, was budgeted at $3.5 billion before construction started in 1975, but recent estimates for the project, which only now is in the final stage toward completion, run to $21 billion, excluding interest.) Social and environmental problems with large hydroelectric projects have become more contentious. Dam reservoirs flood vast tracts of land—roughly 10 times as much land as is needed for comparable coal or solar energy projects—forcing tens of thousands of people from their homes.[58]

Northern aid agencies have recently begun to reconsider their traditional support for large hydroelectric power projects, largely due to public outcry over resulting disruption s. In India, the $11.4 billion Sardar

Sarovar hydropower dam and irrigation project was severely criticized in mid-1992 by an independent review team set up by the World Bank. Among other concerns, the team cited resettlement problems (more than 100,000 people would need to be moved), malaria outbreaks due to the project, and potential impacts to fisheries and people living downstream. The World Bank has so far refused to withdraw its financial support for the project, though Japan's foreign ministry suspended its assistance in 1990. While hydroelectric power will continue to be pursued, countries are likely to design smaller projects with lower economic risk, and fewer environmental and social impacts.[59]

Nuclear power, too, has fallen short of its promise to supply cheap electricity in developing countries—just as it has elsewhere. The Third World accounts for only 6 percent of the world's nuclear generating capacity, with many programs—including those of Argentina, Brazil, and India—over budget, behind schedule, and plagued by technical problems. The Indian nuclear program's operating record is among the world's worst: the country's seven operating reactors at the beginning of 1992 had run just 40 percent of the time, providing less than two percent of India's electricity despite billions of dollars in investment. Taiwan and South Korea have better records, though public opposition has frozen Taiwan's expansion plans since 1985, and, along with rising costs, threatens South Korea's as well. Due to its high cost and complex technology, nuclear power is not a viable option for the vast majority of developing countries.[60]

It is not only the developing countries that face problems with traditional energy supplies such as oil, coal, and hydroelectric power. Most western industrial countries, too, have seen expansion of these fuels slow considerably in recent years. As a result, they have increased their use of natural gas, and more recently, of renewable forms of energy other than hydroelectric power. As technological advances make these resources more economically competitive in the North, they offer new opportunities for the South. Developing countries could select these energy systems now, rather than continue to build infrastructure dependent on the fuels and technologies of the past.[61]

When oil companies operating in developing countries find natural gas (for which export markets are poorly developed) in an exploratory well, they usually cap the well and write the venture off as a tax loss. Yet locally, the gas in these so-called noncommercial wells could provide fuel for cooking, produce electricity, and replace coal and oil in factories and motor vehicles. A natural gas well just one-hundredth the size needed for commercial export could be considered cost-effective for local use, according to Ben Ebenhack, a petroleum engineer at University of Rochester who heads a project to tap previously drilled wells in Africa for local use. The key is building the infrastructure needed to bring natural gas to the large markets waiting for it in developing countries.[62]

Petroleum geologists have already found substantial reserves of natural gas in some 50 developing countries, with many others holding high promise. Most of the identified reserves are located in oil-producing countries such as Algeria, Indonesia, Mexico, Nigeria and Venezuela, many of which have treated gas as a waste byproduct of petroleum production and burned it off without capturing any useful energy. The 21 billion cubic meters of gas Nigeria flared in 1990 could have furnished enough energy to meet all the country's current commercial energy needs, along with those of neighboring Benin, Cameroon, Ghana, Niger, and Togo. But oil exporters are not the only ones who waste natural gas. India burned off more than 5 billion cubic meters of natural gas in 1990, enough energy to save the country nearly $700 million on oil import bills.[63]

Natural gas can be substituted for nearly any energy source used today. According to studies by researchers in India and at the World Bank, compressed natural gas is a cost-effective replacement for diesel and gasoline as a motor vehicle fuel. It can replace oil or coal in industrial processes and be used for cooking, hot water, and space heating in buildings. Propane and other gas liquids, often found along with natural gas, can help reduce urban use of charcoal, wood, and kerosene for cooking.[64]

In developing countries, producing electricity from natural gas, particularly in advanced combined cycle gas turbines, is usually less expensive than using oil- or coal-fired plants, according to the World Bank. Combined-cycle gas turbines, which use jet-engine technology to improve on traditional steam turbines, offer higher conversion efficiency

> "The gas Nigeria flared in 1990 could have furnished enough energy to meet all the country's current commercial energy needs, along with those of neighboring Benin, Cameroon, Ghana, Niger, and Togo."

and lower pollution, as well as lower capital cost. (Combined-cycle gas turbines convert as much as 55 percent of the energy in a unit of gas into electricity; steam plants convert 34 percent on average.)[65]

The revival of natural gas that has taken place in North America and Europe in recent years has also occurred to some extent in the South. Government and private engineers have drawn up plans for vast networks of gas pipelines that would connect developing countries. In Latin America, Argentina could soon be piping gas over the Andes to smog-choked Santiago, Chile. Another network, which received its initial go-ahead in 1992, will feed Bolivian gas to southern Brazil and northern Argentina. And in southeast Asia, Thailand hopes to build a gas grid with neighboring Malaysia and Myanmar (formerly Burma).[66]

Even China is reconsidering natural gas as part of its effort to slow the growth in oil and coal use. In 1986, the government formed a gas research institute, and in early 1992, the government decided to build a pipeline from a large offshore gas field discovered during an unsuccessful search for oil in 1983. Gas commonly accompanies not only oil but also coal, suggesting that China, with its almost limitless coal reserves, is well endowed with natural gas too. One Chinese multiagency group estimates the country's gas resource to be about half as big as its enormous proven coal reserves that would last for well over one hundred years. Recent experience appears to indicate this is correct: in one region in north-central China, every well drilled in the first five months of 1991 struck natural gas.[67]

Developing countries also have abundant supplies of renewable energy resources, such as sunlight, wind, biological sources, and heat from deep within the earth, that are increasingly economical. The past decade has seen dramatic technological improvement in tapping these renewables, trimming costs for solar and wind energy systems, for example, by 66 to 90 percent. Electricity sources such as solar thermal power and photovoltaics could be the least expensive route for developing countries, predicts World Bank economist Dennis Anderson. The higher insolation found at low latitudes, where much of the world's population is concentrated, allows more energy to be produced by each system installed. (Land availability is not a problem: for solar energy to double total current Third World energy consumption, only 0.2 percent of developing-country land area would be needed.) Many renewables are already less expensive than fossil fuels or nuclear power, once social and environmental costs—such as air pollution, resource depletion, and government subsidies—are included.[68]

To take immediate advantage of renewable energy's potential, energy planners need to search out uses that are viable today even without including what they save in pollution costs. Investments in these uses can stimulate development of the technological and business infrastructure and the in-country expertise, both private and public, needed to deploy renewables on a large scale in the future.

Using the sun to heat water is already a cost-effective way for societies to save electricity. Total capital costs for solar hot water are, on average, nearly 25 percent lower than those for electric hot water, when the cost of building power plants is included, according to data collected by the U.S. Office of Technology Assessment. Many developing countries have seen solar industries spring up. Residents of Botswana's capital, Gaborone, have purchased and installed more than 3,000 solar water heaters, displacing nearly 15 percent of the residential electricity demand. Some 30,000 solar hot water heaters have been installed in Colombia, and 17,000 in Kenya. In Jordan, 12 percent of the urban water heating systems are solar.[69]

For grid-connected power supplies, geothermal power plants and new wind generators based on variable-speed turbines can produce electricity at a cost comparable to that from coal-fired power plants. India leads the developing world in wind energy, with 38 megawatts installed by the beginning of 1992. Aided by a Danish joint venture, the country plans to install 1,000 megawatts of domestically manufactured wind turbines by the end of the decade. Excellent wind resources exist in Central America, southern South America, northern Africa, and parts of South Asia, and could provide more than 10 percent of developing-country electricity.[70]

Geothermal energy already plays a major role in some countries; it produced 21 percent of the electricity in the Philippines, 18 percent in El

Salvador, and 11 percent in Kenya in 1990. Yet this resource is abundant—and still largely untapped—in Bolivia, Costa Rica, Ethiopia, India, and Thailand. Another two dozen countries appear to have equally good, though less explored, potential.[71]

Another advanced technology suitable for developing countries is fuel cells. With even higher efficiency—and lower pollution—than gas turbines, fuel cells convert natural gas, biomass, or hydrogen to power and heat. (Unlike conventional power technologies, battery-like fuel cells combine hydrogen-rich gases with oxygen using an electrochemical reaction, instead of combustion, to generate electricity.) Industrial countries, including Japan and the United States, are commercializing new fuel cell technologies that could be useful in developing countries, especially since fuel cells are modular and require less maintenance than standard electric power plants. India has funded a fuel cell demonstration project, though overall investment has so far been low.[72]

Rural areas of developing countries are also ready for renewable energy technologies. In Mexico, the National Solidarity program, a government agency that works to improve economic and social conditions in rural areas, has been deploying wind, photovoltaic, and small hydroelectric power technologies in villages. The program is also converting existing diesel generators to "hybrid" systems that combine diesel with renewable sources. These increase reliability while lowering the amount of fuel that needs to be hauled over long distances. In China, over 110,000 wind turbines are used to provide lighting and power. Small hydroelectric generators supply roughly half of the electricity used in rural parts of the country, where output nearly doubled between 1979 and 1988.[73]

Villagers are also using photovoltaic cells to power lights, radios, and even televisions, needs that are currently met with kerosene lamps and disposable or rechargeable batteries. More than 60,000 photovoltaic lighting units have been installed in recent years in developing countries such as Colombia, the Dominican Republic, Mexico, and Sri Lanka, largely as the result of actions by nongovernmental organizations and private businesses. Africa has experienced a virtual boom in photovoltaic lighting systems since the mid-eighties: Kenya has 10 private companies selling photovoltaics, with as much as a thousand kilowatts installed. Botswana installed 100 kilowatts in 1991 alone. Private businesses in Rwanda, Tanzania, and Zambia are also commercializing photovoltaic lighting kits. Still, at current prices (roughly $500 for a 50-watt system), only a small minority can afford photovoltaics. A survey in the Dominican Republic estimated that just 20 percent of nonelectrified rural households could afford a system even with a seven-year loan. Price declines would increase the number, as could smaller systems. But governments will need to continue their support to ensure complete rural electrification, while focusing not so much on line extension as on dispersed renewables.[74]

Biomass energy sources supply 35 percent of developing-country energy but could contribute more if biomass production were increased and existing agricultural and industrial wastes better utilized. For example, sugar can become more profitable by converting the waste from extracting sugar from cane—called bagasse—into electricity.[75]

Although bagasse is often burned in boilers that use the heat to fuel the sugar extracting process, modern combustion systems can make even better use of bagasse. For each ton of bagasse, a typical sugar mill boiler produces enough steam to produce 15 to 25 kilowatt-hours of electricity while fueling the plant's operation. Modern steam turbines, already in use in some Brazilian plants, can maintain steam production while increasing electricity output eightfold. Gas turbines designed to run on biomass, which are currently being commercialized, could raise electricity output by more than thirtyfold over today's standard boiler (though steam production would fall slightly). If sugar mills burned all their residues using advanced gas turbines, they would meet more than a third of the total current electricity use in developing countries. Gas turbines also could be fired with other agricultural residues, or with forestry wastes currently burned at paper and pulp factories.[76]

Hundreds of millions of hectares of degraded lands could be returned to productivity by planting fast-growing trees and other crops suitable for energy use, according to the U.N. Solar Energy Group for Environment and Development. Any such attempt to boost biofuels production, however, would require major investments by governments and private companies. Past efforts to entice villagers to plant more trees have failed

"If sugar mills burned all their residues using advanced gas turbines, they would meet more than one-third of the total current electricity use in developing countries."

more often than not, particularly if the undertaking is packaged as an energy project. Villagers are often hesitant to plant trees for energy even if they are suffering from fuelwood shortages, partly due to conflicts over land ownership and use. And while Third World women may favor planting because it is they who are responsible for obtaining cooking fuel, it is men who want crops that can be sold in markets instead of those needed at home—and who control land use decisions.[77]

Solutions will require listening to local people, and recognizing and understanding production methods of local farmers. One key is to integrate biomass energy production with a comprehensive agricultural development strategy that produces marketable items, such as timber or pulp. For example, agroforestry techniques, which combine food and wood production, offer a way to boost both food yields and fuelwood and fodder harvests. Research in Kenya and Nigeria has shown that mixing corn and leucaena trees can increase corn production 39 to 83 percent over a field of corn alone, while yielding at least 5 tons of wood per hectare. Even though agroforestry can supply additional fuel for landholders, the poorest of the poor are landless and will continue to face problems meeting their cooking energy needs. Indeed, it is difficult to disentangle rural energy problems, and their solutions, from problems of land ownership, and economic equity.[78]

Together with efficiency improvements on the demand side, an energy system run on renewable energy resources and natural gas has the potential to meet all the new energy needs of developing countries, according to Amulya Reddy and his colleagues at the Indian Institute of Science in Bangalore. Reddy's group crafted a plan that could meet the state of Karnataka's electricity needs in the nineties for only $6 billion of investment, rather than the $17.4 billion a government committee had proposed to spend on large hydroelectric, coal-fired, and nuclear power plants.[79]

Among the new supplies in Reddy's plan are natural gas, solar hot water (to substitute for electric hot water), and more efficient use of existing sugar mill wastes and other biomass. One key to the proposal is redirecting energy investments toward improving the lot of the poor, while de-emphasizing energy-intensive industrialization. Unlike the state's plan, which foresees continuing power shortages despite the enormous investment, Reddy's proposal would electrify all homes in the state and employ more people, while boosting carbon dioxide emissions by only one-fiftieth the amount the government plan envisioned. By combining cost-effective efficiency improvements and new supply options, while better meeting the energy service needs of more people, the plan offers a blueprint for what a sustainable, efficient energy economy would be like.[80]

Energy Policies for Development

In their landmark 1987 study, *Energy for a Sustainable World*, José Goldemberg, Thomas Johansson, Amulya Reddy, and Robert Williams theorized that through efficiency improvements and a shift to modern energy supplies, developing countries could—by consuming just 20 percent more energy per capita than they had in 1980—attain a standard of living similar to that enjoyed by Western European countries in the seventies. By 1990, a 20 percent increase in commercial energy consumption had occurred, yet living standards had scarcely budged.[81]

The problem was not that Goldemberg and his colleagues were wrong, but that more energy does not *automatically* bring improvements in living standards; it has to be combined with policies that can effectively push aside barriers that have impeded global energy reform. The barriers include entrenched special interests with a bias toward large supply-side projects, and consumers who are ignorant of—or unable to finance—efficient alternatives. To overcome these obstacles, governments in both South and North will need to change their long-held assumption that increasing energy supplies—as opposed to the final services energy provides—is their main concern.

The rewards of focusing on energy services can be found in the successes of policies already tried in countries such as China and Brazil. As China learned with its initial progress in the eighties, investing in energy efficiency can yield rewards that developing countries can hardly afford to miss. Yet it will take far more comprehensive changes to turn successful experiments into successful development. The kinds of policies that can

"It is critical that support from the North include not only
the billions of dollars annually provided as foreign assistance,
but also the power of example."

help make this happen include the expunging of destructive subsidies
from energy prices, so that users will have incentives to conserve; shifting emphasis in overall energy planning from building new power-plants and supplies to providing more efficient services to the consumer; and shifting supply-end investment from coal and oil to more benign sources.

To implement these policies will require major institutional changes, from international development agencies, national governments, electrical utilities, individual industries, and nongovernmental organizations advocating sustainable development. It is critical that support from the North include *not only* the billions of dollars annually provided as foreign assistance, but also the power of example. That has already occurred to some degree in the development of more energy-efficient industrial processes and consumer products, but will become far more persuasive when it includes more substantial shifts to renewable, non-polluting energy sources and less energy-intensive lifestyles.

Energy prices are fundamental because if they are too low there is little incentive for the end user to employ energy efficiently, or for anyone to invest in alternative sources. Farmers in India, for example, pay a flat fee or extremely low tariffs for electricity used for irrigation, costing the central government nearly $3 billion a year. The World Bank estimates that developing countries spend more than $49 billion a year subsidizing energy, and has given energy price reform its highest priority. Its expectation is that once prices convey full economic and environmental costs, optimal levels of efficiency will occur.[82]

Experience in industrial countries, however, shows that higher energy prices eliminate only one of the many obstacles to reducing uneconomic energy use. Furthermore, the political reality in developing countries—where inexpensive electricity or diesel fuel is often considered a right—means that it is very difficult to raise prices. To work successfully, substantial price hikes may need to be packaged with direct investments in end-use efficiency, so that the final cost of the energy service—whether cooking, lighting, or pumping water—remains unchanged or actually declines. In India's case, it would be less expensive for the government to retrofit irrigation pumps than to continue funding the enormous subsidies for wasted electricity. Similarly, subsidizing the use of efficient lighting technologies would cost less than providing below-cost electricity to poorer households and building new power plants.[83]

Once end-use efficiency becomes a priority, abundant opportunities for promoting it can be found in industrial equipment, home appliances, buildings, and transportation. Among the programs that can boost efficiency levels are: adopting product efficiency standards; testing, labeling, and publicizing product efficiencies and their benefits to consumers; evaluating industrial projects' or processes' life cycles; and changing the planning process at electric utilities.

Successful examples of such programs can be found throughout the developing world—though none has yet become prevailing practice. But if the key is to make efficiency an overriding objective, Tunisia has taken a significant first step by setting up an audit program—accompanied by low interest loans and favorable tax policies—to improve efficiency in such varied industries as cement, textiles, and hotels.[84]

China provides an example of effective product standards in its requirement of a 30 percent reduction in energy use per square meter in new apartment buildings. Starting in 1993, the standard will be tightened to a 50 percent reduction. Voluntary standards for new buildings, which can lower new building energy consumption by a third, exist in Jamaica, Pakistan, the Philippines, and Thailand. Similar standards can be applied to such products as appliances and motor vehicles.[85]

Brazil provides a successful example of testing and labeling with *PRO-CEL*'s program requiring a standard efficiency test for refrigerators. It has found that simply publicizing the results can push manufacturers to improve performance. Shortly after tests began, average efficiency in new models rose to the level of the previous year's most efficient model. Manufacturers, though not required to build more efficient products, feared that they would lose market share if it became widely known that their models used more energy—and cost more money—to operate.[86]

Perhaps even more effective, in the long run, is to conduct life-cycle evaluations of energy-intensive industries and products before a new facility is built. An Energy Efficiency Impact Statement (EEIS), as proposed by Jayant Sathaye and Ashok Gadgil at the Lawrence Berkeley Laboratory in California, would require builders of new facilities in energy-intensive industries, such as cement and steel, and producers of energy-consuming devices, such as automobiles and appliances, to compare the lifetime energy implications of their facilities or products with those of available alternatives. If a facility requres a higher up-front cost in order to achieve greater energy efficiency and long run cost-effectiveness, the utility company or a government body could provide the additional capital needed.[87]

One way utilities can become major investors in energy efficiency is by adopting a planning process known as integrated resource planning. Originally pioneered by U.S. utilities and regulators, and now spreading to Canada, Japan, and Western Europe, this policy requires power companies considering new generating capacity to compare the proposed expansion with improvements in customer energy efficiency. If efficiency measures prove less expensive, utilities invest money in those, instead of in the new generating capacity. They also have the option of investing in alternative supply options, such as solar hot water heaters, that cost-effectively reduce electricity use. If environmental costs are included in the calculations, efficiency and renewable supplies prove even more economical.[88]

In late 1991, Thailand became one of the first developing countries to adopt a utility-run efficiency program based on the concept of integrated resource planning. By 1997, the country's three power utilities expect to save 225 megawatts of electricity capacity by investing $183 million in efficiency— less than half what it would cost to build an equal amount of new electric generating capacity. Although the current program is expected to reduce the projected increase in electricity use by only 4 percent, studies show that Thailand could push that figure to nearly 25 percent over 10 years.[89]

Examples of efficiency programs, small and experimental as they may be, are more prevalent than alternative supply programs. Yet making this shift is critical, because the environmental consequences of not doing so are unthinkable—and because the conventional resources can't reach everyone. A modest demonstration of what can be achieved is found in programs promoting the distribution of household photo-voltaics to rural areas. The U.S. Department of Energy recently joined with the Brazilian state governments of Pernambuco and Ceara to finance the installation of photovoltaic lighting in some 2,000 homes —the initial phase of a larger project aimed at serving over half a million homes, schools, and health clinics.[90]

Utilities could also play a larger role in implementing the shift to renew-ables, by assisting in the financing of dispersed electrical systems. Some rural banks are already offering small loans, and consumer cooperatives have established revolving credit funds for homeowners to purchase photovoltaic lighting kits. A utility-established fund of $10 million could finance 87,000 household photovoltaic systems in 10 years.[91]

Developing countries can also better utilize domestic resources of natu-ral gas. Since the mid-eighties, India has been working to limit its flaring of natural gas. Even though total flaring still grew by 67 percent between 1985 and 1990, the increase in its use as an energy resource grew faster—by 165 percent—over the same period. In 1991, the World Bank gave India a $450 million loan to capture even more of this wasted resource. The World Bank has increased its support for tapping and building pipelines for domestic supplies of natural gas, which accounted for almost one-third of its energy lending in 1991—up from less than 10 percent in 1990.[92]

Institutional changes needed to effect these shifts begin with internation-al assistance, which unfortunately—the natural gas loans notwithstand-ing—is now doing as much to hinder the needed energy transition in developing countries as to help it. The multilateral development banks, especially the World Bank, play the largest role. While these agencies provide no more than 10 percent of the roughly $60 billion of overall investment made in developing countries' energy sectors each year, they do supply the "seal of approval" that private banks usually require before providing loans.[93]

Judging from their loan portfolios, the multilateral development banks appear to equate energy with expanding electric power, to the virtual

"A utility-established fund of $10 million could finance 87,000 household photovoltaic systems in 10 years."

exclusion of efficiency. Eighty percent of the $54 billion the World Bank has loaned for energy projects since 1948 has been for power supply. Less than 1 percent of the $67 billion loaned for energy by all of the development banks between 1980 and 1990 went to improving end-use energy efficiency, according to research by the International Institute for Energy Conservation. Moreover, the World Bank foresees funding for efficiency remaining at the 1 percent level through 1995.[94]

Changes in priorities are needed, and at the World Bank they may be afoot. Late in 1991, the Bank formed an alternative energy unit in its Asian section to encourage investments in energy efficiency and renewable resource technology. And in July 1992, energy staff presented a new policy paper to the Bank's executive directors laying out the strategy it will follow for future lending to electric utilities. Disturbed by the proposal's reliance on old, failed measures, the Bank's directors refused to approve it, requesting that more emphasis be placed on efficiency. By October 1992, the Bank's directors were pleased enough with the redrafted paper to adopt it as Bank policy.[95]

Still lacking in the new policy paper, however, is the integration of energy policy with broader development goals. The development banks can only fulfill their mandated responsibilities by giving up their destructive practice of simply providing energy supplies, and moving toward a strategy that ties together investments in all sectors. Instead of relying only on politically difficult price hikes to pursue efficiency, for example, the banks could encourage the use of integrated resource planning—as the Asian Development Bank has recently started to do—while investing directly in energy efficiency. One way is to fund national mechanisms that can provide financial assistance to both the public and private sectors.[96]

The Global Environment Facility (GEF), an international fund set up in 1990 and administered by the World Bank, the United Nations Development Programme, and the United Nations Environment Programme, has taken on part of the role of financing efficiency and renewables projects by making grants designed to slow global warming. The GEF has supported some promising initiatives, including a $7 million project for household photovoltaics in Zimbabwe and a $3.3 million project for energy from sugarcane residues in Mauritius. And it is considering several other worthy ideas: installing efficient lighting in Mexico, capturing methane from coal mines in China, financing electricity end-use efficiency improvements in Thailand, developing biomass-fueled gas turbines in Brazil, and promoting a variety of renewable energy technologies in India.[97]

But there are two big limitations with the GEF. First, the GEF's total funds—$1.3 billion for three years, of which only 40 to 50 percent can be energy-related—are not up to the task of reforming energy development. The multilateral development banks lend nine times as much each year on traditional energy projects. Second, by including them in its portfolio, the GEF sustains the false notion that these projects are not economical on their own, but only as a means to reduce emissions of carbon dioxide. The institution's real impact will be felt once the development banks' entire energy loan portfolios follow the pattern of today's GEF grants in emphasizing efficiency and renewables.[98]

The GEF's shortcomings expose a glaring weakness of the United Nations: there is no central U.N. energy agency other than the International Atomic Energy Agency—which promotes the export of nuclear power to developing nations in addition to monitoring nuclear proliferation. To shift the world community toward a sustainable energy economy will require additional international leadership. A new international energy agency could take the lead role in promoting efficient energy systems. Such an agency could be decentralized, incorporating a series of research stations in key regions around the world. These centers could help develop and demonstrate renewable and efficiency technologies, gather and disseminate information, and train technicians and professionals in developing countries. They would be particularly useful to smaller developing countries that do not have the resources of countries such as Brazil or India.[99]

The shortcomings of international agencies are replicated at the country level. Reflecting the World Bank's isolation of energy projects from broader development needs, and the United Nations' lack of an integrated energy agency, individual countries have little coordination among various ministries and the private sector. Yet to ensure that sufficient data on energy needs and use are obtained, and to guarantee that locally based research and development is undertaken, such coordination is essential.[100]

Electric power utilities are in need of major reform. Whether privately- or publicly-owned, the utilities traditionally have been controlled by governments that serve narrow political interests, the effect of which is to keep prices artificially low. While privatizing utilities—as favored by many development and finance experts—can make a difference by increasing capital flow to the electric power sector, it does not necessarily alter the underlying problem of the sector: an overemphasis on building new power plants to the neglect of maintenance and end-use efficiency. Nor does privatization necessarily mean more competition, or even the absence of political interference with the utility's activities. Strong, independent regulatory bodies are needed to assure that efficiency becomes a central concern, that integrated resource planning is adopted, and that private energy producers such as sugar mills can sell excess electricity at fair prices.[101]

If the World Bank can restructure its funding priorities, national governments will likely follow suit—in order to attract their share of international support. Until then, however, many national programs may have to be self-sustaining, applying the principles of a new energy economics in a more limited way.[102]

National programs can be funded by energy taxes or by carbon taxes, which are based on the carbon content of the fuel. In early 1992, the Thai parliament levied a tax on petroleum products and natural gas, equivalent to just over 1 cent per liter of petroleum product, that will provide some $50 to $60 million annually for investments in efficiency and renewables. Over five years, the Thai government expects the tax, along with private businesses, utilities, and a proposed GEF project, to push total investment in energy efficiency and renewables to $500 million. Ghana already funds its independent energy board with a small tax on fossil fuels, and Tunisia originally funded its efficiency program through a modest tax on oil. More general application of carbon taxes would encourage research and investment in efficiency and renewables. The Italian government has recommended that European nations dedicate part of the revenue from a proposed European Community carbon tax to sustainable energy investments in the South.[103]

Many nations are beginning to recognize the connection between energy and environmental stability, which makes such taxes viable. Indeed, the signing of a climate convention by 116 developing countries at the Earth Summit in June, 1992 underscores this. The treaty commits every signatory nation to explore ways to mitigate climate change while preparing inventories of their emissions. Northern countries agreed to adopt policies to limit emissions of greenhouse gases, and through the GEF, to help developing countries slow their emission increases. So far, little new funding has been available, however, and the world community has made no real commitment to help developing countries blaze a new energy path. It is worth noting that Northern countries failed to supply the resources needed to fulfill promises made at the U.N. Conference on New and Renewable Sources of Energy in Nairobi in 1981.[104]

But developing countries need a new energy path for reasons quite apart from threats to the earth's climate. Simply reducing the cost of energy services, as well as the environmental and health costs of air pollution, would allow developing countries to invest in more pressing areas. In Brazil, for example, about 30 percent of children are malnourished and 78 percent do not complete primary school. With the technologies of efficiency and renewables, and with the policies for their dissemination that have already proved effective, Brazil could shift $2 to 3 billion a year out of the power sector and roughly double its funding for nutrition, preventive health care, and water and sanitation programs. Throughout the South, investments in transport and communication systems, health and education infrastructure, water supplies and shelter could be stepped up.[105]

Such investments may not be feasible without the savings generated by a more efficient energy system. Indeed, an energy strategy based on efficiency and low-polluting sources is a cornerstone of the sustainable development process—and necessary if developing countries are to improve living standards for *all* of their citizens. As countries move to this new strategy, their energy economies will shift from obstructing development to enabling it.

Notes

1. Howard Geller, *Efficient Electricity Use: A Development Strategy for Brazil* (Washington, D.C.: American Council for an Energy-Efficient Economy (ACEEE), 1991).

2. Energy consumption is based on British Petroleum (BP), *BP Statistical Review of World Energy* (London: 1992), on United Nations, *1990 Energy Statistics Yearbook* (New York: 1992), and on United Nations, *World Energy Supplies 1950-1974* (New York: 1976).

3. United Nations Development Programme (UNDP), *Human Development Report 1992* (New York: Oxford University Press, 1992); fifty countries is a Worldwatch Institute estimate based on World Bank, *World Development Report 1992* (New York: Oxford University Press, 1992), and on Population Reference Bureau (PRB), "Population Data Sheet 1985," Washington, D.C., 1985; "La Comisión Económica para América Latina Cifra el Número de Indigentes en 183 Millones," *El País*, (Madrid), April 6, 1992.

4. International Monetary Fund, *World Economic Outlook* (Washington, D.C.: May 1992); Geller, *Efficient Electricity Use*; Michael Philips, "Energy Conservation Activities in Latin America and the Caribbean," International Institute for Energy Conservation (IIEC), Washington, D.C., undated.

5. Worldwatch Institute estimate based on Tom Boden, Oak Ridge National Laboratory, Oak Ridge, Tenn., private communication and database, July 28, 1992; Worldwatch Institute based on United Nations, *1990 Energy Statistics Yearbook*, on BP, *BP Statistical Review*, and PRB, on "1991 World Population Data Sheet," Washington, D.C., 1991.

6. Hilary F. French, *Green Revolutions: Environmental Reconstruction in Eastern Europe and the Soviet Union*, Worldwatch Paper 99 (Washington, D.C.: Worldwatch Institute, November 1990).

7. Amulya K.N. Reddy and José Goldemberg, "Energy for the Developing World," *Scientific American*, September 1990.

8. Efficiency's potential savings is based on Mark D. Levine et al., *Energy Efficiency, Developing Nations, and Eastern Europe*, A Report to the U.S. Working Group on Global Energy Efficiency (Washington, D.C.: IIEC, June 1991).

9. Figure 1 is based on BP, *BP Statistical Review*, on United Nations, *World Energy Supplies 1950-1974*, on David Cieslikowski, International Economics/Socio-Economic Data Division, World Bank, private communication and printout, August 24, 1992, on PRB, *1991 World Population Data Sheet*, and on PRB, "World Population Estimates and Projections by Single Years, Less Developed Regions: 1750-2100," Washington, D.C., unpublished printout, March 1992; J.M.O. Scurlock and D.O. Hall, "The Contribution of Biomass to Global Energy Use," *Biomass*, No. 21, 1990; U.S. Congress, Office of Technology Assessment (OTA), *Energy in Developing Countries* (Washington, D.C.: U.S. Government Printing Office (GPO), 1991).

10. Worldwatch Institute based on United Nations, *1990 Energy Statistics Yearbook*, on BP, *BP Statistical Review*, and on PRB, *1991 World Population Data Sheet*; commercial energy refers to oil, natural gas, coal, and electricity sources; it does not include biomass energy supplies such as fuelwood and charcoal which, even though they may be sold in highly developed markets in some countries, lack sufficient data over time for reliable quantification.

11. Income is based on purchasing power parity, from UNDP, *Human Development Report 1992*, and from UNDP, *Human Development Report 1990* (New York: 1990); energy statistics, which include biomass energy use in India and Pakistan, are from D.O. Hall, "Biomass Energy," *Energy Policy*, October 1991, and from United Nations, *1990 Energy Statistics Yearbook*.

12. OTA, *Energy in Developing Countries*; OTA, *Fueling Development: Energy Technologies for Developing Countries* (Washington, D.C.: GPO, 1992).

13. Andrew Hill, "Extracting the Benefits from African Oil," *Financial Times*, May 28, 1992; World Bank, *World Development Report 1992*.

14. C. Rammanohar Reddy et al., "The Debt-Energy Nexus: A Case Study of India," International Energy Initiative, Bangalore, undated; Paul Culbert, "Crisis Hits Huge Market," *Petroleum Economist*, September 1991; David Housego, "India's Economic Growth Rate Slips to 10-Year Low," *Financial Times*, February 28, 1992; International Monetary Fund, *World Economic Outlook* (Washington, D.C.: May 1992).

15. William Keeling, "Oil and Gas Industry: Uncertainty and Expansion," *Financial Times*, June 24, 1992; Mohamad Soerjani, Centre for Research of Human Resources and the Environment, University of Indonesia, "Renewable Energy Resource in Indonesia," presented to Conference on Global Collaboration on a Sustainable Energy Development, Copenhagen, Denmark, April 25-28, 1991; "China Stressing Onshore E&D to Spur Crude Output," *Oil and Gas Journal*, July 29, 1991; China's oil export revenue is in 1991 U.S. dollars and is a Worldwatch Institute estimate based on U.S. Department of Energy (DOE), Energy Information Administration (EIA), *International Energy Annual 1990* (Washington, D.C.: 1992), on DOE, EIA, *Monthly Energy Review May 1992* (Washington, D.C.: 1992), and on BP, *BP Statistical Review*; Lynne Curry, "Doubts Over Role for Foreigners," *Financial Times*, June 16, 1992.

16. Twenty-five percent figure is from Gunter Schramm, "Electric Power in Developing Countries: Status, Problems, Prospects," in *Annual Review of Energy 1990* (Palo Alto, Calif.: Annual Reviews Inc., 1990); percent of new borrowing since the onset of the debt crisis is from John Besant-Jones, senior energy economist, World Bank, Washington, D.C., private communication, August 24, 1992; Organización Latinoamericana de Energía (OLANDE), *La Deuda Externa del Sector Energético de América Latina y el Caribe: Evaluación, Perspectivas y Opciones* (Quito: 1988); tariffs in developing countries are in constant 1986 dollars, from Energy Development Division, *Review of Electricity Tariffs in Developing Countries During the 1980's*, Industry and Energy Department Working Paper, Energy Series Paper No. 32 (Washington, D.C.: World Bank, November 1990); OTA, *Fueling Development*.

17. "World Bank Funds Energy Efficiency Improvement in India," *Global Environment Change Report*, February 14, 1992; Mark D. Levine et al., "China's Energy System: Historical

Evolution, Current Issues, and Prospects," in *Annual Review of Energy and the Environment 1992* (Palo Alto, Calif.: Annual Reviews Inc., in press); World Bank, Industry and Energy Department, "The Bank's Role in the Electric Power Sector: Policies for Effective Institutional, Regulatory, and Financial Reform" (draft), Washington, D.C., April 27, 1992.

48

18. Gita Piramal, "Power Generation a Thorn in the Side of Industrial Growth," *Financial Times*, September 16, 1991.

19. Edwin A. Moore and George Smith, *Capital Expenditures for Electric Power in the Developing Countries in the 1990s*, Industry and Energy Department Working Paper, Energy Series Paper No. 21 (Washington, D.C.: World Bank, February 1990); Schramm, "Electric Power in Developing Countries"; John E. Besant-Jones, senior energy economist, World Bank, "Financing Needs and Issues for Power Sector Development in Developing Countries During the 1990s," presented at the UPDEA-UNIPEDE Symposium on Financing Problems Facing Electrical Power Utilities in Developing Countries, Libreville, Gabon, January 1990.

20. 2.1 billion is from Derek Lovejoy, "Electrification of Rural Areas by Solar PV," *Natural Resources Forum*, May 1992; "Is Rural Electrification Such a Good Idea?" *Energy Economist*, July 1988; Gerald Foley, *Electricity for Rural People* (London: Panos Institute, 1990).

21. P.J. de Groot and D.O. Hall, "Biomass Energy: A New Perspective," prepared for AFREPREN, Third Workshop, University of Botswana, Gaborone, Botswana, December 1989; Gerald Leach and Robin Mearns, *Beyond the Woodfuel Crisis: People, Land and Trees in Africa* (London: Earthscan Publications Ltd., 1988); U.N. Food and Agriculture Organization citation is from OTA, *Energy in Developing Countries*.

22. OTA, *Energy in Developing Countries*; Kenneth Newcombe, "Economic Justification for Rural Afforestation: The Case of Ethiopia," in Gunter Schramm and Jeremy J. Warford, eds., *Environmental Management and Economic Development* (Baltimore, Md.: Johns Hopkins University Press, 1989).

23. World Bank, *World Development Report 1992*; number of deaths from World Health Organization, Division of Epidemiological Surveillance and Health Situation and Trend Assessment, *Global Health Situation and Projections* (Geneva: 1992); Catherine Gathoga, "Indoor Air Pollution," *Stove News*, Foundation for Woodstove Dissemination, Nairobi, May/June 1991.

24. Tim Golden, "Mexico City Emits More Heat Over Pollution," *New York Times*, November 25, 1991; "Mexico City Sets Pollution Record," *Washington Post*, March 17, 1992; OTA, *Energy in Developing Countries*; Bangkok from Victor Mallet, "Third World City, First World Smog," *Financial Times*, March 25, 1992; "Air Pollution Threatens Nairobi Residents," *Daily Nation*, January 9, 1991, in Foreign Broadcast Information Service Daily Report/Africa, Rosslyn, Va., May 7, 1991; "Chilean Capital Chokes on Smog," *Wall Street Journal*, June 13, 1991; Victoria Griffith, "São Paulo Engulfed in a Tide of Pollution," *Financial Times*, May 29, 1991; U.N. Environment Programme and World Health Organization, *Assessment of Urban Air Quality* (Nairobi: Global Environment Monitoring

System, 1988); William U. Chandler et al., "Energy for the Soviet Union, Eastern Europe and China," *Scientific American*, September 1990.

25. Chandler et al., "Energy for the Soviet Union, Eastern Europe and China"; John McBeth, "Heat and Dust," *Far Eastern Economic Review*, September 19, 1991.

26. Carbon numbers are based on Boden, Oak Ridge National Laboratory, private communication and database, and on Intergovernmental Panel on Climate Change (IPCC), *Climate Change: The IPCC Response Strategies* (Washington, D.C.: Island Press, 1991); IPCC, *Climate Change 1992* (Cambridge: Cambridge University Press, 1992); IPCC, *Climate Change: The IPCC Scientific Assessment* (Cambridge: Cambridge University Press, 1990).

27. Paul Lewis, "Danger of Floods Worries Islanders," *New York Times*, May 13, 1992; "Kenya, Other African Nations' Industry Seen Suffering Greatly from Climate Rise," *International Environment Reporter*, May 6, 1992; M.L. Parry et al., eds., *The Potential Socio-Economic Effects of Climate Change: A Summary of Three Regional Assessments* (Nairobi: United Nations Environment Programme, 1991); "China Blames Global Warming for Drought," Reuters, March 24, 1992.

28. BP, *BP Statistical Review*; PRB, "World Population Estimates and Projections"; OTA, *Energy in Developing Countries*; Charles Campbell, Lawrence Berkeley Laboratory, Berkeley, Calif., private communication and printout, June 18, 1992.

29. Worldwatch Institute based on Organisation for Economic Co-operation and Development (OECD), International Energy Agency (IEA), *Energy Policies of IEA Countries: 1990 Review* (Paris: 1991); Levine et al., *Energy Efficiency, Developing Nations, and Eastern Europe*; World Bank, "The Bank's Role in the Electric Power Sector." OCED countries saw energy use grow far slower than economic growth, due in part to improvements in energy efficiency, and to a lesser degree, to structural changes in their industrial bases; for example, OECD countries now focus more on information technologies or "high- tech" products, and less on energy-intensive industries such as aluminium and steel, which increased as a share of developing countries' output (and export to OECD countries). For a discussion of world energy use, structural changes, and intensity over the past twenty years see Steven Meyers and Lee Schipper, "World Energy Use in the 1970s and 1980s: Exploring the Changes," *Annual Review of Energy and the Environment 1992*, (Palo Alto, Calif.: Annual Reviews Inc., in press).

30. United Nations Economic and Social Council, Committee on Energy, "Global Energy Efficiency 21: An Inter-regional Approach," New York, March 6, 1992; OTA, *Fueling Development*.

31. Industrial energy use and Indonesian example from Levine et al., *Energy Efficiency, Developing Nations, and Eastern Europe*; Culbert, "Crisis Hits Huge Market"; C.Y. Wereko-Brobby and John O.C. Nkum, Ministry of Energy, Ghana, "Goals and Means for a Sustainable Energy Development in Africa," presented to Conference on Global Collaboration on a Sustainable Energy Development, Copenhagen, Denmark, April 25-28, 1991.

32. Roger S. Carlsmith et al., "Energy Efficiency: How Far Can We Go?" Oak Ridge National Laboratory, Oak Ridge, Tenn., January 1990; Levine et al., *Energy Efficiency, Developing Nations, and Eastern Europe*; World Resources Institute, *World Resources 1992-1993* (New York: Oxford University Press, 1992); P.M. Nyoike and B.A. Okech, "The Case of Kenya," in M.R. Bhagavan and S. Karekezi, eds., *Energy Management in Africa* (Atlantic Highlands, N.J.: Zed Books Ltd., 1992); Moncef Ben Abdallah, President, Societé Tunisiènne de L'électricité et du Gaz, Tunis, private communication and printout, August 11, 1992.

33. World Bank, "The Bank's Role in the Electric Power Sector"; Besant-Jones, "Financing Needs and Issues for Power Sector Development"; Amalesh Sarkar, "No Sign of Ending Power Crisis: Industries Hard Hit," *Holiday*, Dacca, Bangladesh, May 1, 1992.

34. Ashok Gadgil et al., "Advanced Lighting and Window Technologies for Reducing Electricity Consumption and Peak Demand: Overseas Manufacturing and Marketing Opportunities," Lawrence Berkeley Laboratory, Berkeley, Calif., March 1991.

35. Ibid.; Jayant Sathaye and Ashok Gadgil, "Aggressive Cost-Effective Electricity Conservation," *Energy Policy*, February 1992; Mark D. Levine et al., "Electricity End-Use Efficiency: Experience with Technologies, Markets, and Policies Throughout the World," Lawrence Berkeley Laboratory, Berkeley, Calif., March 1992; Evan Mills, "Efficient Lighting Programs in Europe: Cost-Effectiveness, Consumer Response, and Market Dynamics," *Energy—The International Journal*, forthcoming.

36. Stephen C. Smith, "Industrial Policy in Developing Countries: Reconsidering the Real Sources of Export-Led Growth," Economic Policy Institute, Washington, D.C., 1991; Geller, *Efficient Electricity Use*; Amulya Kumar N. Reddy et al., "A Development-Focused End-Use-Oriented Electricity Scenario for Karnataka," *Economic and Political Weekly.*, April 6, 1991; Lester C. Thurow, *Head to Head: The Coming Economic Battle Among Japan, Europe, and America* (New York: William Morrow and Company, Inc., 1992).

37. Feng Liu et al., "An Overview of Energy Supply and Demand in China," Lawrence Berkeley Laboratory, Berkeley, Calif., May 1992; Hunter Colby, Economic Research Service, United States Department of Agriculture (USDA), private communication, June 17, 1991; USDA, Economic Research Service, *World Agriculture: Trends and Indicators, 1970-1989* (Washington, D.C.: September 1990).

38. U.N. Food and Agriculture Organization (FAO), *FAO Fertilizer Yearbook 1989* (Rome: 1990); Bruce Stone, "Developments in Agricultural Technology," *The China Quarterly*, London, December 1988; Levine et al., *Energy Efficiency, Developing Nations, and Eastern Europe*; Rick Exner, Agronomy Department, Iowa State University, private communication and printout to Peter Weber, Worldwatch Institute, March 11, 1992.

39. Steven Nadel et al., "Opportunities for Improving End-Use Electricity Efficiency in India," ACEEE, Washington, D.C., November 1991; Govinda Rao et al., *The Least-Cost Energy Path for India: Energy Efficient Investments for the Multilateral Development Banks* (Washington, D.C.: IIEC, 1991); P.M. Sadaphal and Bhaskar Natarajan, "Constraints to

Improved Energy Efficiency in Agricultural Pumpsets: The Case of India," *Natural Resources Forum*, August 1992; Sathaye and Gadgil, "Aggressive Cost-Effective Electricity Conservation"; Michael Philips, *The Least Cost Energy Path for Developing Countries: Energy Efficient Investments for the Multilateral Development Banks* (Washington, D.C.: IIEC, 1991); Levine et al., "Electricity End-Use Efficiency."

51

40. Sathaye and Gadgil, "Aggressive Cost-Effective Electricity Conservation"; Liu et al., "An Overview of Energy in China"; Carl Goldstein, "China's Generation Gap," *Far Eastern Economic Review*, June 11, 1992.

41. Liu et al., "An Overview of Energy in China"; Jayant Sathaye and Stephen Tyler, "Transitions in Household Energy Use in Urban China, India, The Philippines, Thailand, and Hong Kong," *Annual Review of Energy and the Environment 1991* (Palo Alto, Calif: Annual Reviews Inc., 1991); Levine et al., "Electricity End-Use Efficiency."

42. Sathaye and Gadgil, "Aggressive Cost-Effective Electricity Conservation."

43. OTA, *Energy in Developing Countries*; Yu Joe Huang, "Potential for and Barriers to Building Energy Conservation in China," *Contemporary Policy Issues* (California State University, Long Beach), July 1990.

44. P.P.S. Gusain, *Cooking Energy in India* (New Delhi: Vikas Publishing House, 1990); Leach and Mearns, *Beyond the Woodfuel Crisis*; Jane Armitage and Gunter Schramm, "Managing the Supply of and Demand for Fuelwood in Africa," in Schramm and Warford, eds., *Environmental Management and Economic Development.*

45. Erik Eckholm, *UNICEF and the Household Fuels Crisis* (New York: UNICEF, 1983); Ogunlade Davidson and Stephen Karekezi, "A New, Environmentally-Sound Energy Strategy for the Development of Sub-Saharan Africa," African Energy Policy Research Network (AFREPREN), Nairobi, January 1992; OTA, *Fueling Development*; "Energy IDEA Award for the Kenya Ceramic Jiko," *Stove News*, July/August 1990; Veena Joshi, "Biomass Burning in India," in Joel S. Levine, ed., *Global Biomass Burning: Atmospheric, Climatic, and Biospheric Implications* (Cambridge, Mass.: The MIT Press, 1991).

46. Levine et al., "China's Energy Situation"; OTA, *Fueling Development*; Motor Vehicle Manufacturers Association, *World Motor Vehicle Data* (Detroit: various years); Mia Layne Birk, IIEC, "The Effects of Transportation Growth on Energy Use, the Environment, and Traffic Congestion: Lessons from Four Case Studies," presented to Transportation Research Board Conference, Washington, D.C., January 12-16, 1992.

47. Mudassar Imran and Philip Barnes, "Energy Demand in the Developing Countries," World Bank Staff Commodity Working Paper Number 25, Washington, D.C., August 1990; Deborah Lynn Bleviss, *The New Oil Crisis and Fuel Economy Technologies: Preparing the Light Transportation Industry for the 1990s* (New York: Quorum Press, 1988); OTA, *Fueling Development*; Rao et al., *The Least-Cost Energy Path for India*; Gregory H. Kats, "Slowing Global Warming and Sustaining Development," *Energy Policy*, January/February 1990; Reena Ramachandran, "Petroleum Conservation—Challenges & Opportunities," *Oil*

Conservation Week Supplement, *Indian Express*, Madras, February 16, 1992.

52

48. Bangkok example is from Yue-Man Yeung, "Great Cities of Eastern Asia," in Mattei Dogan and John D. Kasarda, *The Metropolis Era: A World of Giant Cities*, Vol. 1 (Newbury Park, Calif.: Sage Publications, Inc., 1989), and from "The Future of the Motor Car," *Energy Economist*, January 1992; Marcia D. Lowe, *Alternatives to the Automobile: Transport for Livable Cities*, Worldwatch Paper 98 (Washington, D.C.: Worldwatch Institute, October 1990); Marcia D. Lowe, *Shaping Cities: The Environmental and Human Dimensions*, Worldwatch Paper 105 (Washington, D.C.: Worldwatch Institute, October 1991); Michael Replogle, *Non-Motorized Vehicles in Asian Cities* (Washington, D.C.: World Bank, 1992).

49. Jayant Sathaye and Michael Walsh, "Transportation in Developing Nations: Managing the Institutional and Technological Transition to a Low-Emissions Future," in Irving M. Mintzer, ed., *Confronting Climate Change: Risks, Implications and Responses* (New York: Cambridge University Press, 1992); Jaime Lerner, "The Curitiba Mass Transit System," in Mia Layne Birk and Deborah Lynn Bleviss, eds., *Driving New Directions: Transportation Experiences and Options in Developing Countries* (Washington, D.C.: IIEC, 1991); Kieran Cooke, "A Model for Many Other Countries," *Financial Times*, May 1, 1992; B.W. Ang, "The Use of Traffic Management Systems in Singapore," in Birk and Bleviss, eds., *Driving New Directions*.

50. Rao et al., *The Least Cost Energy Path for India*; Reddy et al., "Debt-Energy Nexus"; Shri K.N. Venkatasubramanian, "Oil Conservation Opportunities in Transport Sector," Oil Conservation Week Supplement, *Indian Express*, Madras, February 16, 1992.

51. Levine et al., *Energy Efficiency, Developing Nations, and Eastern Europe*.

52. Levine et al., "China's Energy System"; Mark D. Levine, program leader, Energy and Environment Division, Lawrence Berkeley Laboratory, Berkeley, Calif., private communication, August 19, 1992.

53. Levine et al., "China's Energy System"; Levine, private communication.

54. Geller, *Efficient Electricity Use*.

55. Worldwatch Institute based on BP, *BP Statistical Review*, on United Nations, *1990 Energy Statistics*, and on Scurlock and Hall, "The Contribution of Biomass to Global Energy Use."

56. BP, *BP Statistical Review*; OTA, *Energy in Developing Countries*; Edward L. Morse, "The Coming Oil Revolution," *Foreign Affairs*, Winter 1990/1991.

57. United Nations, *1990 Energy Statistics Yearbook*; BP, *BP Statistical Review*; Levine et al., "China's Energy System"; Indu Bharti, "Power Generation at Cost of People," *Economic and Political Weekly*, October 20-27, 1990.

58. Moore and Smith, *Capital Expenditures for Electric Power*; Daniel Deudney, *Rivers of Energy: The Hydropower Potential*, Worldwatch Paper 44 (Washington, D.C.: Worldwatch Institute, June 1981); John E. Besant-Jones, *The Future Role of Hydropower in Developing*

Countries, Industry and Energy Department Working Paper, Energy Series Paper No. 15 (Washington, D.C.: World Bank, April 1989); John Dunn, "Water into Juice," *Financial Times*, July 31, 1992; Itaipu costs from Patrick Knight, "Power Rationing in Brazil by 1992?," *Energy Economist*, December 1989, and from Leonard Sklar, International Rivers Network, Berkeley, Calif., private communication, September 22, 1992; Dennis Anderson, *The Energy Industry and Global Warming: New Roles for International Aid* (London: Overseas Development Institute, 1992); Christopher Flavin and Nicholas Lenssen, *Beyond the Petroleum Age: Designing a Solar Economy*, Worldwatch Paper 100 (Washington, D.C.: Worldwatch Institute, December 1990).

59. Bradford Morse and Thomas R. Berger, *Sardar Sarovar*, The Report of the Independent Review (Ottawa: Resource Futures International, 1992); "Japan Halts Cash for Indian Dam," *Daily Yomiuri*, May 31, 1990.

60. Christopher Flavin et al., *The World Nuclear Industry Status Report: 1992* (London, Paris, and Washington D.C.: Greenpeace International, WISE, and Worldwatch Institute, May 1992); Chung-Taek Park, "The Experience of Nuclear Power Development in the Republic of Korea," *Energy Policy*, August 1992.

61. OECD, IEA, *Energy Policies of IEA Countries*; BP, *BP Statistical Review*; Christopher Flavin, "Building a Bridge to Sustainable Energy," in L.R. Brown et al., *State of the World 1992* (New York: W.W. Norton and Company, 1992); Flavin and Lenssen, *Beyond the Petroleum Age*.

62. OTA, *Fueling Development*; Ben Ebenhack, University of Rochester, N.Y., private communication, July 13, 1992.

63. Edwin Moore and Enrique Crousillat, *Prospects for Gas-Fueled Combined-Cycle Power Generation in the Developing Countries*, Industry and Energy Department Working Paper, Energy Series No. 35 (Washington, D.C.: World Bank, May 1991); Nigeria example is a Worldwatch Institute estimate based on United Nations, *1990 Energy Statistics Yearbook*, and on Boden, Oak Ridge National Laboratory, private communication and database; India example is a Worldwatch Institute estimate based on DOE, EIA, *Monthly Energy Review April 1992* (Washington, D.C.: 1992), and on Boden, Oak Ridge National Laboratory, private communication and database.

64. Culbert, "Crisis Hits Huge Market"; Shri K.N. Venkatasubramanian, "Oil Conservation Opportunities in Transport Sector," Oil Conservation Week Supplement, *Indian Express*, Madras, February 16, 1992; René Moreno, Jr., and D.G. Fallen Bailey, *Alternative Transport Fuels from Natural Gas* (Washington, D.C.: World Bank, 1989).

65. Moore and Crousillat, *Prospects for Gas-Fueled Combined-Cycle Power*; Amulya Kumar N. Reddy et al., "Comparative Costs of Electricity Conservation: Centralised and Decentralised Electricity Generation," *Economic and Political Weekly*, June 2, 1990; Andrew Baxter, "By-products of the Jet Age," *Financial Times*, July 31, 1992.

66. Osvaldo F. Canziani, director, Instituto de Estudios e Investigaciones sobre el Medio

53

54

Ambiente, Buenos Aires, private communication to Christopher Flavin, June 6, 1992; José Roberto Moreira, Secretariat of Science and Technology, "Goals and Means of a Sustainable Energy Development in Brazil and South America," Brazil, undated; Bob Williams, "Latin American Petroleum Sector at Crossroads," *Oil & Gas Journal*, July 6, 1992; Government of Thailand, *Thailand National Report to the United Nations Conference on Environment and Development (UNCED)* (Bangkok: June 1992); "A Spat Between Neighbours," *Asiaweek*, April 10, 1992.

67. Liu et al., "An Overview of Energy in China"; P.T. Bangsberg, "Arco's 10-Year Effort Bears Fruit with Pact to Develop Field Off China," *Journal of Commerce*, March 16, 1992; Levine et al., "China's Energy System"; BP, *BP Statistical Review*; "China Stressing Onshore E&D to Spur Crude Output," *Oil and Gas Journal*, July 29, 1991.

68. Thomas B. Johansson et al., "Renewable Fuels and Electricity for a Growing World Economy: Defining and Achieving the Potential," in Thomas B. Johansson et al., eds., *Renewables for Fuel and Electricity* (Washington, D.C.: Island Press, in press); Flavin and Lenssen, *Beyond the Petroleum Age*; Anderson, *The Energy Industry and Global Warming*; Olav Hohmeyer, "Renewables and the Full Costs of Energy," *Energy Policy*, April 1992.

69. OTA, *Fueling Development*; Chris Neme, Memorandum to Mark Levine, Lawrence Berkeley Laboratory, Berkeley, Calif., March 28, 1992; Mario Calderón and Paolo Lugari, Centro Las Gaviotas, Bogota, Colombia, private communication, April 13, 1992; R. Aburas and J.-W. Fromme, "Household Energy Demand in Jordan," *Energy Policy*, July/August 1991; Christopher Hurst, "Establishing New Markets for Mature Energy Equipment in Developing Countries; Experience with Windmills, Hydro-Powered Mills and Solar Water Heaters," *World Development*, Vol. 18, No. 4, 1990.

70. Matthew L. Wald, "High-Tech Windmills Are Sold to the Dutch," *New York Times*, June 16, 1992; Pacific Northwest Laboratory, "World-wide Wind Energy Resource Distribution Estimates," Richland, Wash., 1981; Dennis Elliot, Pacific Northwest Laboratory, Richland, Wash, private communication, May 28, 1991; Neelam Mathews, "Enthusiasm Could Conquer All," *Windpower Monthly*, January 1991; Neelam Mathews, "Bringing Down the Trade Barriers," *Windpower Monthly*, June 1992.

71. United Nations, *1990 Energy Statistics Yearbook*; Ronald DiPippo, "Geothermal Energy: Electricity Generation and Environmental Impact," *Energy Policy*, October 1991.

72. Peter Weber, "Sold on Fuel Cells," *World Watch*, January/February 1992; Koshy Cherail, "Fuel Cells Offer Powerful Hope," *Down To Earth*, New Delhi, July 31, 1992.

73. Jorge M. Huacuz V. and Ana María Martínez L., Electrical Research Institute, Non-Conventional Energy Sources Department, Cuernavaca, Mexico, "Rural Electrification With Renewable Energies in Mexico: Financial, Technical, Social and Institutional Challenges," presented to SADCC Annual Technical Seminar, Swaziland, November 26-28, 1991; Shi Pengfei, "Development and Application of Small Wind Generators in China," *RECs Worldwide*, Folkecenter for Renewable Energy, Hurup Thy, Denmark, March 1992; Liu et al., "An Overview of Energy in China."

74. Lovejoy, "Electrification of Rural Areas by Solar PV"; Mark Hankins, "Home Systems are Fastest Growing Commercial PV Market in Africa," *PV News*, Vol. 11, No. 3; Richard D. Hansen and José G. Martin, "Photovoltaics for Rural Electrification in the Dominican Republic," *Natural Resources Forum*, Vol. 12, No. 2, 1988.

75. Scurlock and Hall, "The Contribution of Biomass to Global Energy Use."

76. Robert H. Williams and Eric D. Larson, Center for Energy and Environmental Studies, Princeton University, "Advanced Gasification-Based Biomass Power Generation and Cogeneration," presented to International Symposium on Environmentally Sound Energy Technologies and their Transfer to Developing Countries and European Economies in Transition, Milan, October 21-25, 1991.

77. U.N. group's recommendations from Johansson et al., "Renewable Fuels and Electricity"; Jodi L. Jacobson, *Gender Bias: Roadblock to Sustainable Development*, Worldwatch Paper 110 (Washington, D.C.: Worldwatch Institute, September 1992).

78. Leach and Mearns, *Beyond the Woodfuel Crisis*; Gerald Leach, "Agroforestry and the Way Out for Africa," in Mohamed Suliman, *Greenhouse Effect and its Impact on Africa* (London: Institute for African Alternatives, 1990); David Brooks and Hartmut Krugmann, "Energy, Environment, and Development: Some Directions for Policy Research," *Energy Policy*, November 1990.

79. Amulya Kumar N. Reddy et al., "A Development-Focused End-Use-Oriented Electricity Scenario for Karnataka," *Economic and Political Weekly*, April 6 and 13, 1991; Reddy and Goldemberg, "Energy for the Developing World."

80. Reddy et al., "A Development-Focused End-Use-Oriented Electricity Scenario for Karnataka"; Reddy and Goldemberg, "Energy for the Developing World."

81. José Goldemberg et al., *Energy for a Sustainable World* (Washington, D.C.: World Resources Institute, 1987); energy consumption increase is based on United Nations, *1990 Energy Statistics Yearbook*, and BP, *BP Statistical Review*; overall, energy consumption per capita rose only 12.2 percent if it is assumed that noncommercial energy consumption per capita remained stable.

82. Rao et al., *The Least Cost Energy Path for India*; Bjorn Larsen and Anwar Shah, "World Fossil Fuels Subsidies and Global Carbon Emissions," World Bank, February 20, 1992; World Bank, Industry and Energy Department, "Energy Efficiency and Conservation in the Developing World: The World Bank's Role," Washington, D.C., March 18, 1992.

83. Michael Grubb et al., *Energy Policies and the Greenhouse Effect*, Volume Two (Brookfield, Vermont: Dartmouth Publishing Company, 1991); Ashok Gadgil and Gilberto De Martino Jannuzzi, "Conservation Potential of Compact Fluorescent Lamps in India and Brazil," Lawrence Berkeley Laboratory, Berkeley, Calif., July 1989.

84. Michael Philips, "Energy Conservation Activities in Africa and Eastern Europe," IIEC,

Washington, D.C., September 1990.

85. Siwei Lang and Yu Joe Huang, "Energy Conservation Standard for Space Heating in Chinese Urban Residential Buildings," *Energy—The International Journal*, forthcoming; Levine et al., "Electricity End-Use Efficiency"; Government of Thailand, *National Report to UNCED*.

86. Geller, *Efficient Energy Use.*

87. Sathaye and Gadgil, "Aggressive Cost-Effective Electricity Conservation."

88. Levine et al., "Electricity End-Use Efficiency"; Sacramento Municipal Utility District, "SMUD Launches Solar Programs," press release, Sacramento, Calif., July 10, 1992.

89. Peter du Pont and Koomchoak Biyaem, "A Walk on the Demand Side: Thailand Launches Its Energy Efficiency Initiatives," in *ACEEE 1992 Summer Study on Energy Efficiency in Buildings* (Berkeley, Calif.: ACEEE, 1992); Mark Cherniack, IIEC, Washington, D.C., private communication, July 14, 1992.

90. DOE, "Agreement Launches Brazilian Solar Rural Electrification Project," press release, June 29, 1992.

91. Lovejoy, "Electrification of Rural Areas by Solar PV."

92. Gas flaring based on Boden, ORNL, private communication and database; World Bank, Energy Development Division, "FY91 Annual Sector Review: Energy," Washington, D.C., October 23, 1991.

93. Besant-Jones, World Bank, private communication; Levine et al., *Energy Efficiency, Developing Nations, and Eastern Europe.*

94. World Bank, *Annual Report 1992* (Washington, D.C.: 1992); African Development Bank, *1990 Annual Report* (Abidjan: 1991); Asian Development Bank, *Annual Report 1991* (Manila: 1992); Inter-American Development Bank, *1991 Annual Report* (Washington, D.C.: 1992); Philips, *The Least Cost Energy Path*; World Bank, Energy Development Division, "FY91 Annual Sector Review: Energy."

95. "World Bank Board Critical of Staff Policy Papers," *Banknote*, IIEC, September 1992; Glenn Prickett, Natural Resources Defense Council, Washington, D.C., private communications, September 30, 1992; E. Patrick Coady, executive director to the World Bank, presentation to the Environment and Energy Workgroup, Society for International Development, Washington, D.C., October 15, 1992.

96. Philips, *The Least Cost Energy Path*; Cherniack, private communication.

97. Global Environment Facility (GEF), "Report by the Chairman to the April 1992

Participants' Meeting," Washington, D.C., March 1992; GEF, "A Selection of Projects from the First Three Tranches," Working Paper Series Number II, Washington, D.C., June 1992; GEF, "Mauritius Sugar Bio-Energy Technology Project," Washington, D.C., January 1992.

98. GEF, "Report by the Chairman"; Philips, *The Least Cost Energy Path.*

99. Anderson, *The Energy Industry and Global Warming.*

100. Amulya K.N. Reddy, "Barriers to Improvements in Energy Efficiency," Lawrence Berkeley Laboratory, Berkeley, Calif., October 1991; Philips, *The Least Cost Energy Path*; Brooks and Krugmann, "Energy, Environment, and Development."

101. Rajendra K. Pachauri, "Population and Energy," in Vasant Gowarker, ed., *Science, Population and Development: An Exploration of Interconnectivities and Action Possibilities in India* (New Delhi: Unmesh Communications, 1992); Adilson de Oliveira and Gordon MacKerron, "Is the World Bank Approach to Structural Reform Supported by Experience of Electricity Privatization in the UK?," *Energy Policy*, February 1992.

102. Reddy, "Barriers to Improvements in Energy Efficiency"; Philips, *The Least Cost Energy Path.*

103. du Pont and Biyaem, "A Walk on the Demand Side"; Cherniack, private communication; Davidson and Karekezi, "A New, Environmentally-Sound Energy Strategy"; Philips, "Energy Conservation Activities in Africa"; Anderson, *The Energy Industry and Global Warming*; Haig Simonian, "Gap Needs Bridging," *Financial Times*, October 20, 1991.

104. UNCED Secretariat, "154 Signatures on Climate Convention in Rio," press release, June 14, 1992; Michael Grubb, "The Climate Change Convention: An Assessment," *International Environment Reporter*, August 12, 1992; Committee on the Development and Utilization of New and Renewable Sources of Energy, "Solar Energy: A Strategy in Support of Environment and Development," Report to the Secretary-General, United Nations, New York, February 13, 1992.

105. Brazil example is from Geller, *Efficient Electricity Use*; Davidson and Karekezi, "A New, Environmentally-Sound Energy Strategy."

NICHOLAS LENSSEN is a Research Associate with the Worldwatch Institute, and coauthor of the organization's annual report, *State of the World*. He is a graduate of Dartmouth College, where he received a degree in geography. From 1984 through 1987, he worked with the U.S. Peace Corps in Ecuador.

THE WORLDWATCH PAPER SERIES

No. of
Copies

_____ 92. **Poverty and the Environment: Reversing the Downward Spiral** by Alan B. Durning.

_____ 93. **Water for Agriculture: Facing the Limits** by Sandra Postel.

_____ 94. **Clearing the Air: A Global Agenda** by Hilary F. French.

_____ 95. **Apartheid's Environmental Toll** by Alan B. Durning.

_____ 96. **Swords Into Plowshares: Converting to a Peace Economy** by Michael Renner.

_____ 97. **The Global Politics of Abortion** by Jodi L. Jacobson.

_____ 98. **Alternatives to the Automobile: Transport for Livable Cities** by Marcia D. Lowe.

_____ 99. **Green Revolutions: Environmental Reconstruction in Eastern Europe and the Soviet Union** by Hilary F. French.

_____100. **Beyond the Petroleum Age: Designing a Solar Economy** by Christopher Flavin and Nicholas Lenssen.

_____101. **Discarding the Throwaway Society** by John E. Young.

_____102. **Women's Reproductive Health: The Silent Emergency** by Jodi L. Jacobson.

_____103. **Taking Stock: Animal Farming and the Environment** by Alan B. Durning and Holly B. Brough.

_____104. **Jobs in a Sustainable Economy** by Michael Renner.

_____105. **Shaping Cities: The Environmental and Human Dimensions** by Marcia D. Lowe.

_____106. **Nuclear Waste: The Problem That Won't Go Away** by Nicholas Lenssen.

_____107. **After the Earth Summit: The Future of Environmental Governance** by Hilary F. French.

_____108. **Life Support: Conserving Biological Diversity** by John C. Ryan.

_____109. **Mining the Earth** by John E. Young.

_____110. **Gender Bias: Roadblock to Sustainable Development** by Jodi L. Jacobson.

_____111. **Empowering Development: The New Energy Equation** by Nicholas Lenssen.

_____ **Total Copies**

☐ **Single Copy: $5.00**
☐ **Bulk Copies (any combination of titles)**
    ☐ 2–5: $4.00 each   ☐ 6–20: $3.00 each   ☐ 21 or more: $2.00 each

☐ **Membership in the Worldwatch Library: $25.00 (international airmail $40.00)**
The paperback edition of our 250-page "annual physical of the planet," *State of the World 1993,* plus all Worldwatch Papers released during the calendar year.

☐ **Subscription to *World Watch* Magazine: $15.00 (international airmail $30.00)**
Stay abreast of global environmental trends and issues with our award-winning, eminently readable bimonthly magazine.

No postage required on prepaid orders. Minimum $3 postage and handling charge on unpaid orders.

Make check payable to Worldwatch Institute
1776 Massachusetts Avenue, N.W., Washington, D.C. 20036-1904 USA

Enclosed is my check for U.S. $_____

_____

name   **daytime phone #**

_____

address

_____

city   **state**   **zip/country**

five dollars

Worldwatch Institute
1776 Massachusetts Avenue, N.W.
Washington, D.C. 20036 USA